Backyard Homesteading

*An Essential Homestead Guide to Growing
Food, Raising Chickens, and Creating a
Mini-Farm for Self Sufficiency and Profit*

Contents

Part 1: Backyard Homestead

The Ultimate Homesteading Guide to Growing Your Own Food, Raising Chickens, and Mini-Farming for Self Sufficiency and Profit

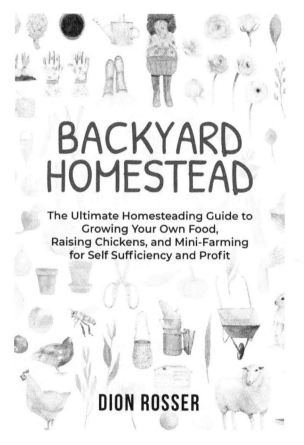

Introduction

Backyard Homestead introduces the principles of homesteading with hands-on methods that beginners can use right away. This complete guide walks you through all the great reasons to homestead, how to get started, what challenges you should plan for, and much more.

Don't miss the opportunity to learn about homesteading in an easy to understand and practical manner. These methods are fool-proof and have been used by households operating in a self-sufficient manner for decades or even centuries. Many people now remember that only a few generations ago most households relied on their own garden, raised chickens, and engaged within their community. This self-sufficiency is once again becoming a more normal situation. It doesn't matter if you have a small backyard or acres of property, you can get started with the basics and learn how to plan and develop your homestead as your skills develop even further.

Keep reading to discover how you can implement modern techniques to master age-old methods of cultivating your land and a lifestyle that fits your values.

Chapter 1: Reasons for Starting a Homestead

It doesn't matter if you've always had an interest in gardening or if changes in the economy and personal security have made you consider going back to homesteading. Anyone can homestead if they have a little property and a touch of patience. Okay, there are certain things you will need, but the most important thing to get you started is your reason you want to do it. With one good reason, you can drastically change the course of day-to-day happenings of your life.

One or more of these reasons may have spurred your interest in homesteading. Exploring the reasons you're interested in homesteading can help you set out your goals and plans before you jump right in.

Better Health

It's generally common knowledge that eating organic fruit and vegetables is better for your health. That belief leads most people to the conclusion that if they grew their own fruit and vegetables, it would be a better option than shopping in stores. There is some truth to homegrown fruit and vegetables having more health benefits, better nutrient density, and fewer additives. However, it's important to take a rational or scientific approach to the association between farming and health.

Let's roll through some quick facts that can help you determine if this is the right reason for you to start homesteading:

• Homegrown food does have more nutrients; vegetables and fruit begin to lose nutritional value within 24 hours of harvest, ergo, fresher food means more nutritional benefits.

• No presence of genetic modification – at this time there is no conclusive evidence that GMOs are harmful. However, they aren't natural and are likely not beneficial.

• No harmful pesticides, waxes, or other chemicals. You have control over unwanted additives to your food.

Along with homegrown fruit and vegetables, you can make better health choices through homesteading by raising chickens, goats, and cattle. Essentially what you're doing regarding your health is taking control over the processes applied to your food before it gets to your table. You don't have to worry about pesticides unless you're choosing which pesticides to use in your garden. Many homesteaders use only natural remedies such as vinegar or "garden buddies" – specific herbs that can drive out pests. This level of control can allow for a much healthier lifestyle.

For example, did you know there are at least twenty-one common fruits and vegetables, including avocados and apples, that are treated with wax before they arrive at grocery stores? When it comes to health, it goes without saying that organic is better. Homegrown is also better, as long as you take measures to ensure that health and safety are your priority.

Educational Experience

How much do you know about the food you buy? Do you know the life cycle of a plant? What about the incubation period for chicken eggs?

Homesteading leads you to ask questions that would never cross your mind. It makes you think about growth and development as a cycle rather than a journey with a destination. Many homesteaders cite the challenge or the educational experience as the reason for starting. Many others report they started homesteading because they had children and wanted to teach them the value of growing their own food and being self-sustaining.

If you're not sure about including the educational experience as part of your plan for making goals, consider the educational gaps that will become present as you start homesteading. Whether you want the education to be part of your experience or not, it will be necessary to have a successful garden and crop rotations. Most of our basic survival and living skills have been lost because of the massive development of our food product supply chain. On the part of economics, it's made tons of jobs and helped build a stable element in our day-to-day society. But when it comes to developing yourself as a person, it's evident that most of us lack the necessary skills to live independently.

Connection with Your Food

It's easy to go through a Jack-in-the-Box drive-through or order out from Slaters 50/50 and forget that your burger once had a face. Now there's nothing terribly wrong about eating meat. However, it is important to understand where your food comes from and what role it plays in your life. Many who include chickens or milking cows or goats in their homesteading plan find that they do connect with the animals. It can make dealing with the loss of an animal, or consumption, difficult to handle at certain times. That deeper connection to food can not only make you more appreciative of the contributions these animals make in our daily lives, but also the role of food waste in society. People who want to connect with their food better may also aspire to lead zero-waste lives or better manage their connection with the Earth and life around us.

Homesteading can drastically change how you view food and food sources, both financially and emotionally. Very few people use this reason as their primary decision-maker to start homesteading. And, of those people, you'll likely encounter more vegetarians or vegans as our modern lifestyles have disconnected us from the correlation between animals and meat.

No matter what your choice of diet, having a deeper connection with your food can change that relationship exponentially. Where you may once have taken for granted the overstocked shelves in a superstore, you suddenly begin to realize the lifecycles and planning that go into raising chickens or managing crop rotations. Even if this isn't your primary reason for homesteading, it's likely to be a byproduct of the homesteading experience.

Better At-Home Dining Experience

At-home dining is something that we can all enjoy. It's highly likely that you and most people you know can cook better food than that you normally experience when dining out. When you look at the most commonly frequented restaurant chains, including Texas Roadhouse and The Cheesecake Factory, most items on their menu are not made in-house. Rather they're prepared in a manufacturing facility, packaged, and then usually steamed-in-a-bag or fried on site.

But having said this, how can homesteading lead to better at-home dining? Three primary factors can affect your cooking and dining experiences at home.

First, fresh tomatoes, eggs, herbs, and everything else always taste better. Food grown or cultivated at home always tastes better. Honestly, it might be entirely psychological if it weren't for the chemicals and treatments that grocery-store food goes through. We've already covered how homegrown food is better. But when it comes to preparing and eating food, there's the matter of freshness. Even if you shop at higher-end markets such as Sprouts or Whole Foods, that produce spends hours or even days on trucks traveling in the open air and under the sun; the food simply can't stay fresh in those conditions. When you're homesteading, the food goes from the vine or plant into your kitchen or refrigerator. You can't get fresher than that!

Second, you have greater control over possible contaminants and your ingredient management. Homesteading can open your eyes. One homesteader explained that she started after learning that about 95% of grocery store products contained corn or corn byproducts. Well, some things such as cornbread mix were understandable, other things such as fruit snacks were not. What's troublesome with revelations like this is that most of the food in our grocery stores is not as nutrient-dense as it should be because of the presence of corn byproducts. It was one of the things that changed her mind about buying from stores, and it's helped others to see the many advantages of homesteading. For many, tidbits of information like this are the first step in the educational experience that comes with deciding to become a homesteader.

Finally, we have a psychological element. You're more likely to take more care in preparing your ingredients when you are the one who put in all the hard work of growing those vegetables or milking that cow. If you nurtured those little tomatoes from sprouts to sun-ripened globes of goodness, you're going to be much more upset if your sauce doesn't turn out well. So, the natural solution is to handle your ingredients well and develop your cooking skills.

Freedom from Corporate Product-Supply Chain

There are many social and economic effects that we experience every day because of the corporate product-supply chain, better known as the U.S. Food System. Now this reason for homesteading isn't exclusively about living off the grid or "fighting the system." Homesteading can be beneficial for your local economy and strengthen you, and your neighbor's, role in the nation's security and stability.

The U.S. Food System has become a series of tournaments where mass-producers compete for space in the most frequented grocery stores across the nation. That's a drastically different picture from what was happening a few decades ago, where farmers contributed to local or regionalized grocery stores. It's extremely difficult for local farmers to rival major agriculture manufacturers, although there is an increased awareness of local farmers and the need to buy locally.

Dependence on the system that feeds into your major grocery stores perpetuates the difficulties that local farmers and local economies experience. The freedom from this structure can allow you to eat seasonally and better manage or simplify your life when it comes to food. When was the last time you saw anything but the general vegetables in your store? You'd be hard-pressed to find rhubarb in most common grocery stores, even when it is in season.

Security from Panic Buying and Compromised Economic Situations

In times of crisis, people rush to the grocery store. The brief history of panic buying includes the 13 days leading up to the Cuban-missile crisis, the 1973 and 1979 oil crises, Y2K, the 2008 economic recession, and of course, 2020's COVID-19. That's six major instances of panic buying in less than 60 years. These examples are limited to survival-style panic buying, meaning that we didn't include the New Coke panic buying or the resurgence of Crystal Pepsi panic buying. Then you have seasonal over-buying of Frankenberry, Count Chocula, Peeps, and other limited-time items.

Aside from panic buying, you have compromised economic situations. There are times when the agriculture or food service industries are compromised by events that don't largely affect other industries. For example, the Mad Cow outbreak in 2003 made prices of beef skyrocket across multiple countries, although the United Kingdom was the most affected. It economically impacted countries that didn't actually have cases of Mad Cow. The agriculture industry doesn't have to experience the full-scale effect of the Dust Bowl all over again to create an economic issue. One bad crop rotation in America, especially a corn crop, can lead to extensive issues mostly inside grocery stores. People aren't losing their jobs or going out and buying tons of toilet paper at a time, but they are facing extreme price hikes because of the scarcity of primary staples of our diet.

When you're homesteading, you don't have to worry about that. If you raise and slaughter your own cattle or chickens, changes in meat prices aren't really an issue. You might be more concerned about the availability of alfalfa. You don't have to worry about panic buying as much when you know that most of your household supplies come from materials that you cultivate. It's possible to use natural ingredients grown at home for a great variety of other household products, including cleaners, skin and body care products, and more. Many people turn to homesteading following a panic-buying experience because it's traumatic. Panic-buying leads to real shortages with a serious impact on the people who didn't go out and panic-buy. The only ones purely unaffected by panic buying are the people who cultivate their own food.

Understand Your Reasons for Homesteading

Your personal reasons for homesteading are probably good reasons, even if they're not listed above. What you have to consider when you start planning, or even just thinking about the possibility of homesteading, is how you'll lay out your goals. Homesteading, especially backyard homesteading, calls for a very specific formula for success. It is made up of two parts planning, one part hard work, and one part consistent maintenance. Your reasons will direct your goals and put you on the path to make a plan that will allow you to execute those goals. It's critical that you clearly understand your reasons for homesteading.

Some reasons that people list for their desire to homestead are actually the products of a homesteading lifestyle. It's like saying, "I want to open a company to earn money." You don't have to own a business to make money, and you don't have to homestead to simplify your life or better control your diet choices. However, those are huge payoffs that come with growing your own food or raising your own chickens. Let's break down why simplification and controlling diet are not reasons for homesteading but instead natural benefits.

Simplifying your life is not an immediate thing you'll experience through homesteading. In fact, you'll largely complicate your life for the first year. You'll worry over plant health, ground rotations, planting methods, and seeding. You'll add numerous daily tasks to your schedule such as watering plants, feeding chickens, gathering eggs, weeding planters or troughs, and more. After your first year, when you have a schedule and know what works, you will probably have a much simpler life. You won't worry about what to get at the grocery store, at least in terms of vegetables. You won't need to worry so much about menu planning, or to dine out as often, and certainly won't have to worry so much over your budget. If you thought that simplifying your life was your reason for homesteading, you might consider prioritizing one of the reasons above to help manage your goals and planning.

Better control over your diet is another direct product of homesteading, but if you're looking to use it as your reason for homesteading, you might reconsider and use better health as your primary reason or goal. Often when people start homesteading to limit or restrict their diet, they give up or are too restrictive. What you plant dictates what you eat, and if you're only planting veggies to restrict your diet, you may create so many restrictions that it's not sustainable for your health. No matter what your reason is for homesteading, sustainability is the foundation of a successful homestead and a healthy diet. If you only grow zucchini and tomato, it's not sustainable. Diversity, nutrition, and the season will greatly impact your homestead planning, and if controlling your diet is your primary reason, you may need to reframe your mindset on your diet in order to match the goal of health, wellness, and home-growing your produce and possibly raising animals.

Your reason for homesteading is largely purposeful to you and will direct how you approach your plan and execution of creating your backyard homestead. In the chapters to follow, you'll see how your reasons for homesteading will play a part in building self-sustainability, plotting out your mini-farm, building coops, preparing your kitchen, and much more. Very few people start out understanding the full impact of homesteading on day-to-day events, and homesteading is an ongoing learning experience even for those who have been at it for years. Take your reason and make it a lifestyle for homesteading success.

Chapter 2: Six Top Things to Consider When Planning a Homestead

When you're working on putting a homestead together, you'll need to break the giant project down into tiny manageable tasks. Planning can drastically change the level of success you experience on your homestead, and you'll need to start planning before you even buy your first seeds. Not only will you learn how to plan out each section of your homestead, but you'll develop planning skills that you'll need to maintain your homestead, share your experiences, and connect with more people through the homesteading community.

So, where should you start? You'll need to decide what elements of homesteading you'll choose to participate in, and then what other elements will naturally come from those choices.

Will you:

- Grow vegetables and roots?
- Grow herbs?
- Plant fruit trees?
- Keep chickens?

- Raise chickens for multiple purposes?
- Keep goats for ground maintenance and milking?
- Raise cattle for the diary?
- Raise cattle for meat or slaughter?
- Preserve food through canning, dehydrating, freezing, or in a root cellar?
- Make your own goods such as jams, preserves, and dairy products?

As you go through each of these ask yourself these questions:

- Do I have the space available?
- Is my climate appropriate?
- What equipment or furnishings will I need?

Understanding your answers to these questions doesn't mean that you can or can't homestead. Virtually anyone can build more sustainability into their households through growing food and upcycling. However, it can drastically change what homesteading means for you. Use this list to set your priorities and understand your current limitations.

To help explore your possibilities and potential limitations, we'll go through the planning element in four parts.

1. Land use
2. Household needs and limits
3. Location restrictions
4. Skills and abilities

Land Use

Remember that the idea of a homestead is for self-sufficient living. You only need to cultivate the land in a way that supports your style of living and a healthy diet. If you're not a big, fruit-that-grows-on-trees kind of person, then focus on berries and vine fruit instead. Each of these things will help feed the other. For example, from the garden, you can grow vegetables to feed the chickens and goats. The goats (with some rotation) will help keep your ground in good condition and create fertilizer for your garden, and they also produce milk. A homestead is truly the place where nothing goes to waste, and what happens when you have actual, "I can't use this," waste? It goes on the compost pile, to create better soil for next year.

Consider how you answered the questions above and how important each factor is to you when it comes to homesteading. If you don't have access to certain resources, then you may need to reprioritize your homestead goals temporarily.

Household Needs and Limits

How much self-sufficiency can you build into your household? With an open mind, a touch of creativity, and some ingenuity, you may be able to restrict or even eliminate your regular shopping habits. Of course, no one judges you for the occasional purchase! But imagine if you could grow all of your vegetables right in your backyard and likely have a much more diverse offering than you can find in your local grocery store? You can have all the eggs you could need for the year with just a few chickens. With a goat or cow, you can have a regular flow of milk. When you expand and bring in rabbits, you have access to some of the more expensive meat varieties without the expensive price tag.

"Happiness belongs to the self-sufficient." - Aristotle

So, knowing that your household could be almost completely self-sufficient, even to the extent of creating your own cleaners and skincare products, the question becomes once again about space and storage.

How and where will you store or hold your goods? Some items can last for years on a shelf, for example, jarred peaches. Those peaches, if jarred properly, are something you can hold onto for a long time. But your tree won't stop producing more peaches, so what will you do with those? Plants produce at their own rate and you can either:

- Eat at the rate of plant production (a tough task unless you have a large family)
- Pickle, jar, can, or store in a controlled environment
- Sell or share your excess

Part of learning how to homestead is learning how to do things the way they were done not so long ago. It does seem archaic to have a root cellar or to spend your Sundays pickling eggs, but honestly, these are the tried and true methods for preservation. But what if you genuinely don't have space to store everything properly? Just because something can be jarred or canned doesn't mean that you can continue storing the new influx for weeks or months at a time.

The good news is that homesteading allows you to share the wealth because, for most residences trying to sustain a household, there's always excess. If you genuinely run out of space to store your harvest, any part of it, then share. Sell fresh items at a local farmer's market or give away items to your neighbors and friends. It's a great way to become friendlier with neighbors, and a dozen eggs can earn a lot of forgiveness for the strife over your chicken's noise level.

We've managed to cover how you can be entirely self-sufficient and that if you run into storage space limitations, you have other options. But what about your access to equipment and supplies or even the financial means to get started? Many people enjoy saying, "I can't," but they don't look at the situation enough to generate any creative solutions. Yes, you may need pallets, wood, saws (which are expensive), seeds (which can become expensive), and help in the way of manual labor. However, there are excellent solutions to all of these challenges. If you're facing some trouble over these challenges to getting started, here are a few creative and often low-budget solutions:

- Pallets are an excellent wood source and can often be purchased from trucking yards for less than $3 per pallet.
- Start with cheaper crops – you don't need a rare breed of cucumber; you just need cucumbers.
- Rent equipment from hardware stores
- Borrow equipment from friends or family (and if you want to, call them in to help!)

Homesteading is hard, and one of the most difficult things that people experience is the new sensation of replacing, "I can't" with "I'll figure it out." However, it is a very empowering mindset change. Nothing is stopping you except space limitations, and when you hit those limitations, you can share or sell your goods until you re-stabilize your goods rotation and have space again.

Location and Financial Restrictions

Location and financial restrictions do pose some unique challenges, and it's difficult to overcome these issues sometimes. When it comes to location, your city or county may have restrictions on noise levels, the number of animals you can have on a property, or where you can dig or move utilities throughout your property. Additionally, if you're renting your home, you may feel like you're even more restricted.

Let's cover the issues with the local government first. The biggest trouble is running water and power through your yard. Usually, water isn't an issue because you will be creating nothing more complex than you would expect with a sprinkler system, and many households have those across the country. But electricity is a big obstacle. If you need to connect a water heater and lights out in your chicken coop but need to run a line to make that happen, you may need certain permits from your city. They don't want you to create a fire hazard or not adhere to local electrical or building codes, and they don't want you accidentally digging into an already existing utility line. These are pretty reasonable concerns, so if you need to move access to electricity, talk to your city or county office about your property and your options. Know that extension cords are not a long-term solution, and they can be dangerous, especially if you have animals such as chickens or goats that can get to them and fray the wires.

Now, if you're renting, you likely do have some restrictions. However, you can create collapsible structures. For example, most property owners aren't opposed to garden planters, hanging gardens, or trough gardens as long as they don't do lasting damage to the yard. And you might choose to use hydroponics in your garage, or similar alternatives to avoid disrupting the yard. Renting is complicated and varies from person to person. If you're renting now, you might consider making it a priority to put some money aside to purchase land or participate in a local community garden where you can still garden in the dirt and reap the benefits. Finally, you might explore the options for loans or grants that help small farms and homesteads get started. You can find more resources on funding options in Chapter Nine.

Then there are animal zoning regulations that apply to both homeowners and renters. These regulations vary from county to county and often have nothing to do with the noise levels, but the space and the humane raising of animals. A general guide, although your area may vary, gives a limit of two cattle, four sheep or goats, or two pigs on a 1/2-acre property. These guidelines typically use an "or", not an "and", clause meaning that you can have two cattle or two pigs but not both. There are also restrictions on the total volume of animals, and specific restrictions on how many chickens, dogs, and even cats or rabbits you can keep. You can find your local regulations by searching online "animal zoning regulations + (name of your county)" or visiting your county regulations office.

These challenges bring us back to the notion that some planning and prioritizing can help you overcome many obstacles. But, with location and financial obstacles, there may be some things you have to accept. If your city restricts noise levels and doesn't allow residential chickens, then you may not be able to keep chickens given your current residence. Animal zoning regulations are a struggle. Here are some creative options to overcome or work around common obstacles with location and financial limitations:

• Segment your homestead in a way that fits your budget. For example, build a chicken coop with your savings in January, then build trough gardens with your tax return in April. In May, purchase your seeds, then in July buy your fruit trees, and in September set up your goat pen.

- There are resources for funding (grants and loans) available through the government on a federal level, state level, and depending on your location, possibly even through your county or city offices.

- If you can't garden at home, get involved with a community garden, or help to set one up near your neighborhood.

- Prioritize your animals or use a smaller blend to adhere to animal zoning – for example, having one goat, one cow, and four chickens may be doable versus having two cows and 10 chickens. Explore different ways to stay within local regulations by prioritizing your short and long-term homesteading goals.

- Problems with running electrical underground: consult with a local electrician to determine if you can weather/child/animal protect a wall-mounted electrical line.

Skills and Abilities

You know that with some creative planning, you can get around most limitations, but what about the things you just don't know how to do? Well, homesteading is an ongoing learning process. Even after you've run a homestead for five or even ten years, you'll learn new things. During the first three years of homesteading, however, you'll face a steep learning curve. Fortunately, if you don't learn something the first time, it doesn't mean that your entire mini farm fails.

You may need to learn how to build things with your hands, such as planter boxes or a chicken coop. You will need to learn how to care for and maintain plants, and maybe some small animals. However, you may not have realized you must also learn how to plan around weather changes, and even develop your decision-making skills. Imagine that a heatwave blows through your town, and your poor chickens and rabbits have no relief but the natural shade, which is hardly any relief at all. Do you spray them down with water? Do you freeze water bottles and let them huddle around the sweet icy relief? You will build skills and abilities as you go, but to get started, you'll want to learn about gardening best practices and basic building skills. You might also want to hone your skills on upcycling, as it gets the creative homesteader juices flowing.

How to Set Realistic Goals for Your Homestead Plan

Nothing is more motivating than a worthy goal, and with homesteading, you need to think both long-term and short-term. Your goals will help guide you as you build your homestead and develop it over the first few years. Here are a few common sample goals for new homesteaders. Use them to help get you thinking about your goals!

Long-Term Goals

- Purchase a lot for a large homestead – 10 years + goal
- Add steers to cow herd – 5-year goal
- Raise rabbits for meat – 3-year goal
- Raise chickens – 1-year goal

Short-Term Goals

- Plan seasonal garden and crop rotations on a calendar – 1-week goal
- Build three box gardens – 1-month goal
- Plant first seasonal garden – 2-month goal
- Obtain all materials necessary to build a chicken coop and "run" – 3-month goal
- Make space for a dairy cow – 4-month goal
- Start preserving first harvest – 4-month goal
- Purchase dairy cow – 5-month goal
- Remove eggs from your shopping list – 6-month goal

Prioritizing your Goals

Always prioritize what is most important to you. Some things are easier to manage at different times of the year, so consider that. But in addition to your listed goals, you might create a list of skills you need to help set your priorities. Using the short-term goal list above as an example, the first goal was to plan out the season's crops on a calendar. That means you would take time researching when certain plants thrive. The second goal is building planters, and you might do that research at the same time as when you're planning your calendar because many of the resources may be the same.

When you're planning your homestead, it might seem like every big milestone is weeks or months away. But with a map of your yard, after you've planned out the layout, you can start looking for things to do right now. You might prepare your yard, buy the lumber, or get a feel for the airflow and sun-to-shade ratio on your property. There's always something to be done on a homestead, and you can start right now as long as you have a basic plan.

Chapter 3: Making a Plan for Your Mini Farm

When you're planning out your homestead, you'll quickly become a master project manager, goal setter, and resource navigator. Possibly the most difficult part of homesteading is the initial planning. As Brett Brian famously said, "Farming is a profession of hope," and all the planning that you're doing right now is laying down your seeds of hope.

Setting Realistic Goals as You Plan

As you go through the planning, you'll want to keep realistic goal setting in mind. That doesn't mean lowering your expectations. It simply means that your goals are within the realms of reason. Agriculture and homesteading works largely on natural laws and a strong reliance on Mother Nature. If you plant seeds today, you certainly can't eat tomatoes tomorrow. Setting realistic goals means including the average period of time required for any given activity or development and ensuring that the goal is measurable. Here are two examples of a good goal and a less than realistic goal:

Unrealistic Goal: Make planter boxes and plant seeds.

Realistic goal: Buy materials and make planter boxes this weekend, set up mulch and water systems for seeding next weekend.

The second goal has two primary differences. First, it gives a time frame to both steps. Second, it lists the necessary tasks within the goal. You can use the 'steppingstone' goal system or the 'SMART' goal system or any other option that fits your personality; just make sure that you give everything a time frame or due date and understand all the tasks involved with completing that goal.

What Happens When You Don't Have a Plan?

Disaster; absolute and utter disaster. While there are many things in your life that you can simply "wing" or go with the flow, homesteading is not one of them. You must plan for the season you're in, and the season ahead of you. You must plan how you'll store your harvest for months to come, and what you might do to improve efficiency in day-to-day chores. If you can't see the forest for the trees, you'll be lost.

Imagine if you went out and bought chickens today but didn't have a coop. Chickens have a lot of natural predators, and chicks are tiny. A hawk could easily come and clear out your new flock, or a coyote could get them overnight. Or, imagine if you bought a trough garden today and threw some general seeds in it, but then realized that you have a two-week vacation coming up, and didn't make plans for anyone to care for the seedlings.

Fortunately, you can plan around everything. Plan around your vacations, your weekend plans, your time spent at work, or how the weather changes with the seasons. Unfortunately, however, there will be times when you're planning doesn't quite work out. Things change, the unexpected happens. But still, it is always better to have a plan and adapt later, rather than to not have a plan at all.

"By failing to prepare, you're preparing to fail." - Benjamin Franklin

Quick Considerations for Your Layout Plan

In chapter two, we covered a lot of elements that you should carefully consider before you even start planning. We won't go through them all again, but instead, we will leave you with a shortlist that you can refer back to quickly throughout the remainder of this chapter:

- Water access
- Sunshine exposure and shade coverage
- Land grade (will you need to level land?)
- Trees and existing structures that cannot be moved

Mapping Your Mini-Farm Layout

Everyone has a restriction when it comes to their land; even if you have acres and acres of property, there are restrictions. However, a limitation of land or space doesn't mean that you can't homestead. In fact, many people successfully homestead on less than an acre, or even only a quarter of an acre of land. Even if you only have a small square of a backyard, you have a lot of opportunities to build up your homestead.

To start planning how you will use your land, follow these steps:

1. Measure the space in your yard and start creating a map.

2. Mark out how far away you have access to electricity through outlets, noting where you can occasionally use extension cords.

3. Mark out on your "map" any large, already existing, structures (play centers, gazebos, sheds, etc.)

4. Mark on your map where you already have water access

5. If you have a patio, mark where the cement ends, and the yard begins

Then consider how much space you'll give to each aspect of your homestead. Think about how you answered the questions above and take a look at your map. Then use a few average sizes of structures to help you plan out your space.

Typical Average Sizes to Keep in Mind

- Most trough gardens are 4' x 8' or 4' x 12' – Trough gardens can also have multiple tiers or levels, a bonus space-saving tip! (May need a lot of sun.)

- Chicken coops – Most chicken coops (comfortable for up to 10 chickens) are 4' x 4' or 4' x 8' with a chicken run of about 4' x 8'. (Requires sun and shade.)

- Goat pens and pasture – Goats need a lot of space. You can opt for a night-time "pen" of about 4' x 8' with a top, or a full pen of 10' x 6'. (Requires sun and shade.)

- Milking cow pasture – 15 square feet per cow, or

- Fruit trees – each tree needs about 20' x 20'

- Berry bushes – plant about 2' to 2.5' apart

- Other elements such as alfalfa crops, corn, and compost piles you can size to your needs.

When making your map, know that you don't need a lot of backyard space unless you have children. Even then, children love playing around animals, trees, and a garden. You may set aside some space for a lawn, sitting area, or play area, but you can, without a doubt, create a self-sustaining homestead on a quarter of an acre. The benefit of having trough gardens and flexible fencing designs is that you can move some things around as needed. Now your crops, chicken coop, trees, and bushes aren't so easily moved. Just be sure that when you go through your initial planning, you're putting everything in an area that meets all of its needs so that you're not constantly moving things around. It is always easier to over-plan and do things right than to have to try to move a chicken coop three months later when you realize it has no natural shade in the daytime. However, some of your household needs and limitations may affect your land use as well. For example, if you need chickens, then without a doubt, you're dedicating a fair amount of space to a chicken coop and run. Additionally, if you don't have the means to preserve all of your products, then you may want to scale down your first-year goals to ensure you're not overrun with unused produce, dairy, and eggs.

Sample Backyard Homestead Layouts

As you go through the layouts, keep in mind that you can level land, remove trees, and add trees or awnings for shade. If you find yourself thinking, "That won't work," you can try again.

1/4-Acre Homesteads

On a quarter of an acre, you can manage a small herb garden as close to your backdoor as possible, and if you have space next to that, you can plant two or three berry bushes. They're generally not large and can do with partial sun and partial shade. Then, within your yard, you should be able to manage three planter boxes, or three stretches of 4'x4', 4'x6', or 4'x8' gardens. With your property, you should still have room for either two fruit trees, a goat pasture, or a single cow pasture. Finally, you should have room for a 4'x4' chicken coop with an additional 4' for a chicken run.

That kind of setup should not consume so much of your yard that you can't have room for your kids to play outside or a nice sitting area. The idea, however, is to keep your plants as close to the backdoor as possible so that you can place your animals further out. The chicken area may be the only thing you want to change as it can be nice to have the coop near an outdoor outlet near the side of the house. Most houses have one or two weather-safe outlets facing into the backyard either on the back wall or on the sidewall. Those can be helpful for lamps and water heaters.

An alternative setup is to divide your grassy area into two pastures, one containing the chicken coop. Both pastures can contain fruit trees. Then, on the cement area of your back yard, you can have smaller planter boxes to fit within your patio area.

1/3-Acre Homesteads

With a one-third acre, you have a lot more space, but still, need to be efficient with your planning. One option is to segment off an area four feet deep along the entire backstretch of your property line and use that for a pasture either for a goat, cow, or pig. On the other side of the pasture, you could put your chicken coop with rabbit cages on the other side of the chickens. Remember that rabbits are more skittish than chickens, so it's best to give them some space from larger animals no matter how good-natured they are.

In the largest and sunniest area of your yard, place your garden. You can opt to dig into the ground or use planter boxes. Just ensure that you have enough space to walk, weed, and harvest. Again, keep your herb garden as close to the house as possible. Consider placing trees or berry bushes on the least utilized section of the yard. This style of setup should leave you with enough space for a swing-set or play area and maybe even a fire pit. You can have your homestead and enjoy your backyard too!

1/2-Acre Homesteads

Now you have a lot of room! A half of an acre might not seem like much, but you're running a mini-farm, not a mass production farm, and you have more than enough space to enjoy every element of homesteading.

Start by planning your orchard. The good thing about orchards is that they don't need level ground. If possible, keep your chicken coop as close to your orchard as possible, it's great natural shade and keeps temperatures down during summer months. Along with the poultry area, you can keep rabbits as well. Do try to keep goats away from trees, if possible, as they like to eat the tree itself, not just the leaves.

Then, use a larger portion of your level and sunny area for vegetables, alfalfa, and berries. You can grow in rows or even fence them off from one another. Then you should have a rectangular area that you can segment off for a goat or cow pasture. Try to keep these further away from the house, and you should have room for a large compost bin near the pastures.

As always, herb gardens are best nearer the house, but when you have gardens this large, you might consider planting your herbs within your garden to keep out pests. Many herbs can deter bugs that could otherwise devastate your garden, and with a larger garden, you need all the help you can get.

1-Acre Homesteads

Oh, the things you can do with a whole acre of property! Start with your herb garden, or again, utilize your herbs for low-level pest control in your garden area. You may also aim to separate your seasonal garden from your perennial garden. Using planter boxes is still nice as it can reduce the amount of time you spent stooped over, but it is also nice to work right into the ground.

Then carve out a back corner of your property for compost. With this much land, you can get away with an open-air compost pile so long as it doesn't bother your neighbors. Then, you can have an orchard area. If your orchard is properly fenced, your orchard area can double as an open-air chicken run; place the chicken coop inside the orchard. Chickens love to run around and explore, and most breeds don't fly well, but if you notice one or two of them getting out, you can simply clip off the final corner feathers of their wings. If you're raising chicken for meat, this is exceptionally good.

With your orchard and garden areas set up, you can focus on your pastures. Again, within the confines of your animal zoning, you should be able to get away with sheep, goats, and a cow or two. Sheep and goats do well together but don't keep them with the cows because they can exchange unwanted gastrointestinal parasites that do serious damage to your cows.

Regardless of Size

Before moving on there is one last note when it comes to planning out your homestead: nearly everything is possible. Some successful homesteaders work with 1/8th of an acre of land. They use hanging gardens, have tiny chicken coops for two or three chickens, and use a compost bin. Don't buy into the idea that you need a massive amount of space to homestead. You may need to start small and then grow later.

Near the end of this book, we have a chapter on growing your homestead, and that may mean growing physically or adding new elements. But there are almost always ways to get started with what you have now. It's one of the great lessons of homesteading; there is always a way to "make do."

Making the Most of Your Land

There are many ways that you can get multi-functional use from your land, although again, it takes a bit of planning to execute effectively. Your pastures, coops, orchards, berries, and gardens don't all need separate spaces. Additionally, you can make the flow easier for harvesting and maintenance, ensuring that you use your land in the most efficient manner possible.

For example, one of the sample backyard homestead layouts explained above made the orchard part of the cow pasture. That was possible because the trees needed about 20 square feet each, and the cows needed a pasture of 10'x10' meaning that you could have two trees along with your two cows. Cows and orchards work well together because unlike goats, your cows shouldn't try to eat the bark or low hanging branches of your trees (although they do like to rub up against them.)

Another common pairing is to keep sheep and goats together. Their temperaments mix well, although you might consider castrating males to discourage crossbreeding, as it rarely works out well.

We can go over some examples of making your flow easier for working and harvesting. Initially, many people choose to keep their herb garden on a kitchen windowsill as it makes the herbs easily accessible and doesn't take up outdoor space. Another option is to arrange your orchard or berry bushes on the side of the yard that doesn't get much foot traffic, as you'll harvest from these plants less often. You might also keep your pastures or pens close to your compost pile. Finally, consider how you feed and rotate through your harvest. Do you give the bunnies fresh vegetables? Why not sit them right near your garden?

Sitting down and mapping out your plan can quickly seem overwhelming, but it's important to understand that even with a 1/4-acre of land, there are a thousand and one ways to plan your homestead successfully. Don't get overwhelmed and remember that if necessary, you can change the layout. Although changing the layout is a hassle, it's possible.

Chapter 4: Selecting Your Seed Stock and Cultivars

Your seeds and cultivars will primarily dictate what you gain from your homestead. The decisions you make here will determine most of the plant-based portion of your diet. You can choose from a greater variety of seed and plant options than you might find in the store. However, many factors go into making these decisions. The first step is to decide if you'll use seeds, cultivars, or a blend of both.

What is the Difference between Seeds and Cultivars?

Seeds are pretty straightforward. They are seeds. Some are extremely expensive, and it's generally best that you get them from a reputable seed supplier rather than the paper packets you can find in stores. If paper packets have worked for you in the past, then stick with what works. Otherwise, you might jump online to explore the options in seeds after you decide what exactly you want to grow.

Cultivars, however, are an entirely different approach to planting. Cultivars come from the plant itself; they're a portion of a tree, a trimming from a plant, or something similar. You can do it with all kinds of fruit and vegetables; for example, you can grow a new celery plant from the heart of an old one. However, cultivars with the exclusion of fruit trees, are often a disappointment. Typically, these plants can't sustain their life or produce high-value vegetables. Additionally, if you use the seeds from these plants, they often produce the same lackluster result. Now cultivars can happen naturally. However, many of the cultivars that you'll come across on the market today are plants that are patented and licensed, meaning that you're not purchasing a natural product of a plant, but something created for this very purpose. Now some of these purposes make sense. One example is the Burning Bush shrub, which through cultivation, making it a cultivar, has become more compact.

Perhaps the best analogy to explain the difference between seeds and cultivars is dog breeding. Seeds are your very natural one-dog-meets-another-and-they-have-puppies scenario. Cultivars, or modern cultivars, are more like puppy mills. Through excessive inbreeding and controlled genes, you produce an organism with very specific traits. Unfortunately, the traits have become so specific that they are unstable.

As a new homesteader, you may choose to use a blend of seeds and cultivars. It may be easier to get started with cultivars even if their lifespan may not be exceptionally fruitful or long. There may also be the challenge of deciding exactly which seeds you want right away; using cultivars for one season at a time can help delay that decision.

Building a Garden for Your Season and Climate

It would be wonderful if the sky were the limit when it came to gardening, but everyone has to take into consideration the season and the climate. As a new homesteader, it is reasonable to dread the upcoming vicious winter months that tend to wipe out gardens. However, there are a wide variety of very hardy winter vegetables and crop options.

We've tried to break your options down within these restrictions as much as possible, so it's easy to refer back to this chapter when you're planning your seasonal crops. Now, although there many different climates, because of new technology and improved agriculture methods, it is possible to grow things that wouldn't normally do well in your environment. While you may not be able to grow an avocado tree in the New Mexico desert with ease, it could be possible.

Spring

- Asparagus
- Avocados
- Broccoli
- Cabbage
- Carrots
- Celery
- Collard greens
- Radishes
- Rhubarb
- Strawberries
- Swiss Chard
- Onions
- Mushrooms
- Lettuce
- Garlic
- Peas

Summer

- Okra
- Lima beans
- Raspberries
- Strawberries
- Summer squash
- Tomatillo
- Tomatoes
- Zucchini
- Eggplant
- Corn
- Garlic
- Cherries

- Celery
- Carrots
- Cantaloupe
- Blueberries
- Blackberries
- Bell peppers

Fall: (Nearly All Summer Crops are Suitable for Fall as Well)

- Potatoes
- Pears
- Pumpkins
- Rutabagas
- Sweet potatoes
- Yams
- Swiss chard
- Cranberries
- Ginger

Winter

- Carrots
- Celery
- Collard greens
- Kale
- Leeks
- Onions
- Pumpkins
- Swiss chard
- Winter squash
- Turnips

10 Perennial Vegetables to Grow Year-Round

Some plants are perennial, which means that they can produce year-round, and often they can produce for more than two years. These are often the vegetables and grains that you find in most modern diets.

So, what can you grow almost anywhere and year-round? These ten vegetables will grow at any time of year when properly cared for:

• Tomatoes – They can grow for years but can't survive a harsh winter (with full ground freeze). You can bring your tomatoes indoors during the winter or use weather-safe external heating elements.

• Peppers – They survive almost every climate but may need to come inside during the winter.

• Eggplants – Vegetables that will grow year-round. However, they are often treated as an annual plant.

• Okra – Can grow as high as 7-feet tall

• Chayote squash – A vine vegetable that goes dormant through the winter but produces from the beginning of spring to the end of fall.

• Horseradish – A rooted plant that usually gets harvested in winter for curing bad colds, something you can use year-round.

• Onions – Some varieties will grow year-round, including the perennial leeks, Egyptian walking onions, and perlite onions.

• Artichokes – Specifically the Jerusalem artichoke, it's a very hardy plant similar to potatoes. But be careful, they can grow and spread like wildfire; they may need their own container so as not to choke out your other plants.

• Radicchio – An all-star in salads and similar to cabbage, this plant wants sun and will reappear every spring, although it goes dormant in fall and winter.

• Kale – Grows well in hot and cold weather and is generally referred to as a super crop. Not to mention that it's packed with nutrients.

Keep in mind that you may not want to grow these year-round vegetables, or you may need to segment your perennial garden from your seasonal garden.

Plants That Only Grow in Certain Climates

It might be surprising to learn that tomatoes weren't originally native to Italy. In fact, they made their way to Italy by way of a

Spanish vessel that had come from Peru. Until tomatoes made their way to Europe, they only thrived in the Peruvian climate. Now they've been bred to be so hardy that they are one of the go-to plants for new gardeners.

There are some plants, however, that are just too difficult to grow; mostly humid or tropical plants. Things like Asian greens, Chinese cabbages, Bok Choi, and tropical lettuce such as salad mallow, and even taro only do well in hot weather and high humidity.

What Can Flowers Do for Your Garden?

Flower gardens are beautiful, and vegetable gardens are useful, but why keep them apart? Few beginner growers realize the full potential of having flowers within your garden. They have exceptional benefits and can help you better develop your green thumb as they do require a bit more attention than most beginning vegetables.

Flowers serve three primary purposes in a garden:

● Promote pollination – resulting in higher yields and longer plant life

● Act as a decoy to deter pests – aphids and other pests often prefer flowers over vegetables.

● Attract predatory insects – bring in the insects that prey on aphids and other garden pests.

Flowers attract more pollinating insects which can not only help fertilize the flowers, but also tomatoes, beans, zucchini, peas, and any crops that rely on pollination. And they act as a bit of a sacrifice. If given the option between your tomatoes and cosmos, those aphids will flock to the flowers. Then you have predatory insects such as wasps and hoverflies that will eat aphids and nearly any other insect they can prey upon.

Know Your Seed Varieties

Typically, when discussing seed varieties, people are talking about the different plant life they have on their property. However, there are different types of actual seeds that you may purchase for planting. This is the information that you'll see on seed labels, and it can be a bit difficult to understand at first. But knowing what you're buying is pretty important.

Open-Pollinated

These seeds are from plants that naturally bred in a field or general growth environment. Usually, these strains are much more stable because they've been bred generation after generation, and they should produce plants similar to their parent.

Heirloom

These are open-pollinated plants that were bred for specific traits for a minimum of 50 years. It is why heirloom tomato varieties are often so easy to distinguish. Heirlooms are well-known for their hardiness, which can make them a better choice for people who haven't gardened previously.

Hybrid

Hybrids use purposeful breeding to crossbreed two different plants or even two different species. However, hybrids are altered in the field through a "natural" means of pollination and cultivation. Hybrids do carry the risk of being sterile, and they may not breed "true," meaning that they may not produce exactly what the breeder or grower intended.

GMO

GMO is exactly what you probably suspect it is; these are Genetically Modified Organisms. They are made in a laboratory and are typically patented, resulting in brand name labels, and it is a challenge to find these on the home growing market.

Cell-Fusion CMS Treated

A method of genetically altering the seed through a type of GMO process. The modified trait is eliminated from future genetic possibilities, which means that it comes from a GMO, but it does not share the same genetic modifications as the parent plant.

Why is knowing all of this about plants and seeds important? As noted earlier, heirlooms can be easier to grow. Additionally, hybrids may carry a higher-nutrient density or a better disposition for your climate. Although these are details that you may not want to concern yourself with during your first season, it is something to consider as you look into seed buying more deeply.

Use Your Personal Preferences to Plan Your Seed Stock

Tomatoes are generally easy to grow in a wide variety of climates and can grow year-round if handled correctly. But, if you don't like tomatoes, don't grow them. Think about what you eat now and what you would like to introduce into your diet. Do try to be as diverse as possible and give some thought to planting different varieties within the same family, such as gold potatoes and red potatoes.

But in the end, you don't want to have a garden full of things you don't like to eat. Part of your garden's role in your homestead is to cultivate self-sustainability within your household, meaning fewer trips to the grocery store or dining out. Keep in mind that your garden may not just include the vegetables that take center stage on your dinner plate. Many grow kales purely for juicing or smoothies, and others grow peppers to season and spice up their cooking. Even the flowers mentioned earlier are largely edible. For example, clover and marigold are both edible but have very different roles on a table.

Use your personal taste to lay out the foundation of what you would like to see in your garden. Then determine what seasons those choices thrive in, and if there are additional plants you can add to your garden to support their growth and expand the variety of options you have available in your kitchen.

It might seem like all this research into seeds is unnecessary, or maybe overkill. However, you'll save yourself a ton of time and effort if you put your energy into choosing plants that you enjoy eating, that can be harvested in an easy-to-manage rotation, and that will hold up better to disease.

Chapter 5: How to Select Chickens and Build a Coop

Chickens can be an extremely valuable element of any homestead. They not only produce eggs, but they also provide fertilizer and make excellent pets. However, like any other animal, they entail quite a bit of work. It's not as easy as sticking them into the coop and checking for eggs every few days. They require daily care, can become sick, and may need special help in either cold or hot seasons. The presence of eggs can also attract unwanted creatures onto your homesteads such as snakes, skunks, opossums, rats, raccoons, and crows.

Although there is the risk of pests, chickens are well worth the investment and position within your homestead. What you'll need to do is plan how you will house them, feed them, and keep their environments clean. How you'll go about planning these things typically won't change based on what type of chickens you have.

Selecting the Right Chickens for Your Homestead

When choosing your chickens, you'll want to consider the volume of egg production, how resilient they are, and their abilities to promote your homestead in other ways. For example, some chickens are excellent for pest control. Although they may invite in unwanted small mammals, they can control bugs on the property. Other chickens are better for meat production while still offering egg production.

The three primary types of chickens we'll explore here are the heritage breeds, the alternative breeds, and "momma hen" breeds.

Heritage Breeds

Typically, your heritage breeds are the more common chicken breeds. Heritage breeds include the White Leghorn, the Rhode Island Red, and the Black Australorp. They do have some basic differences, especially when it comes to production and temperament. They're easy to distinguish by color and size.

Usually, homesteaders will have at least one heritage variety, and many beginners will have all three. The purpose of starting only with heritage breeds is that they're all pretty hardy. However, everyone has their preferences, and bringing in all three breeds is like a bit of an experiment to see what options fit you best.

White Leghorns

White Leghorns, as in the famous cartoon chicken Foghorn Leghorn, are possibly the most common type of chicken. They are your stereotypical white-feathered chicken with a yellow beak and red comb or craw. They lay between 250 and 300 eggs per year, and their egg-laying largely depends on climate and security. They are perhaps the most consistent chicken breed when it comes to egg production.

White Leghorns typically have a docile demeanor. Their temperament is calm, but largely they like to be left alone; they're not the type of chicken you buy for animal support or to serve as a pet. White Leghorn roosters can become aggressive.

Rhode Island Reds

These chickens lay the coveted brown eggs and regularly produce about 280 eggs per year. They're a great choice for first-time homesteaders because they are a very resilient breed. They stand up to both cold and warm weather well. You can easily recognize them for their pretty red feathers, and they're generally smaller and stouter than Leghorns.

These chickens are extremely well tempered, although again, the roosters can become aggressive. Rhode Island Reds are also great foragers and actively go after insects. If you're looking for something with a better temper than a Leghorn and are willing to sacrifice some egg production, you might choose Rhode Island Reds.

Black Australorps

Australorps will lay between 200 and 240 eggs in a year, a slightly lower egg count than the other two heritage breeds. You can easily distinguish them by their black feathers with almost green highlights. They're generally pretty and can get rather large. They are rather gentle but can be timid as they tend to scare more easily than most other breeds.

Initially, the Black Australorps came from an English breed called Black Orpingtons to increase egg production without increasing the size of the bird or decreasing the quality of the meat. If you're looking for a bird for meat and egg production, then you should definitely consider the Black Australorp.

Alternative Breeds

These aren't heritage breeds, and often they're mixes of one or more breeds bred together for a specific purpose. Often the purpose is meat quality or hardiness when it comes to holding up against some weather conditions. These breeds are generally hardier, calmer, and more economical in that their production isn't shifted by their space. Alternative breeds are also great for urban settings. If you don't have a lot of land, these are a top choice.

California Whites

These chickens are egg production machines! They typically put out about 300 eggs per year and have very little fluctuation in production from season to season. They are winter hardy and do well with confinement. The confinement element is one thing that makes them so desirable in small environments. They lay white eggs. And they are smaller than your Leghorn varieties, although they're a mixed breed between a Leghorn and various other breeds.

Finally, the Whites are very quiet and docile. If you're worried about the noise level of your chickens, you might need some California Whites. If you know that your neighbors would have problems with louder flocks, a California White flock would be best. What deters most people who are new to chicken raising from California White's is that the hens are "broody," which means they want to sit on their eggs. You may see some aggression when you go to collect eggs, or have hens outright refusing to leave the nesting boxes.

Cinnamon Queens

Red, stout, and a high-production brown egg-layer. Because of its hybrid breeding, the Cinnamon Queen develops and begins laying quickly. They have feathers that go from dark red across the top of their body to light red underneath. Their legs are primarily yellow, and their beak is well-colored. They lay between 250 and 300 eggs per year, but they do stop producing earlier in life, which means that after a few years you may have a lot of retired hens that aren't producing but still draining homestead resources such as feed, water, and space.

Additionally, there is quite a bit of concern about Cinnamon Queen's health troubles. Many report that they die very early on, and it can be extremely unpleasant. These statements aren't a shock as Cinnamon Queens are a "designer" red-gene chicken, which is the equivalent to a designer dog breed that has been overbred into health problems.

Red and Black Sex Links

Red and Black Sex Link chickens are not inherently an alternative breed but a collection of smaller breeds. These breeds, including Cinnamon Queen and many others, were bred with the specific purpose of getting a high egg production. That means that they often lay well above 300 eggs per year. But, don't be fooled, higher egg production often comes with various health issues, bad temperaments, and high activity. They also don't do well at all in cold weather.

If you have a lot of land, then the high activity isn't a problem, but it is discouraging, particularly for new chicken farmers, when you don't know how to help the potential health problems. Of course, many of these chickens live full, disease-free lives. It's just one of the many possibilities to go over before you make your decision on which chickens to purchase.

Raising Chickens for Slaughter

Most of the chickens you would choose for egg-laying aren't the best options for meat production. The best options for meat production are the Cornish hens and the Rock chickens. Additionally, you might choose Freedom Rangers, which generally grow slower, or the Plymouth Barred Rock, which generally produces more meat than the Cornish or Rock varieties. However, the meat is not of the same quality.

There is a significant difference between chickens meant for meat production and chickens meant for egg production. When it comes to chickens meant for meat production, your heritage breeds are the Cornish hens, rock hens, and Freedom Rangers. These chickens are well known for their friendly demeanor and natural tendency to explore. They want to roam around; they don't want to be stuck in a small coop for the entirety of their lives.

When considering the choice of raising chickens for meat, you'll need to go through the following steps:

- Can you give these chickens a good life and a humane death?
- Can you financially provide for chickens that may not give back to the homestead for months?
- Will you raise heritage breeds or Cornish cross hybrids?

The first issue mentioned here can help you determine what type of chickens you choose if you're raising them for slaughter. If you cannot give chickens a good life and a humane death, then you should generally avoid raising chickens for meat. However, if you're worried about giving them a good life for the duration of a few years, that may change your perspective. Most chickens for meat production are humanely slaughtered before they reach eight months in age. You don't have to give this chick a home for two or three years. You simply have to give them food, water, and shelter for a few months. Then you'll need to worry about how you can humanely slaughter them. There are a few different tools to help farmers with this, and as you develop your homestead, you'll quickly learn that animal death is one of the elements you'll have to face on occasion.

When it comes to the financial elements, there are a few misconceptions. When you first start raising chickens, you'll come across one of the few times in homesteading where it is more expensive to have the homestead than to go to the grocery store. Your initial investment to raise chickens will involve a coop, access to water, feeders, and nesting boxes. Then you'll have your recurring expenses for the chickens, including water and chicken feed. If you calculate the cost through your first few months raising chickens for meat production, it will definitely seem as though spending $2.99 per pound for chicken thigh meat at the grocery store is cheaper. However, as you get into the ebb and flow of raising chickens for meat production, the initial costs and the recurring costs will level out.

So, will you raise heritage chickens or Cornish cross hybrids for meat production? Many people are encouraged to use heritage chickens for several reasons. First, their natural curiosity to explore and run around can help with pest control and dirt. Second, because they are running around, their meat often has a more natural taste. Finally, they often make great companion pets for other chickens and goats.

Cornish cross hybrids, however, are a different story. These chickens have been bred to sit, stand, and eat. They are not interested in interacting with other animals, or in exploring the yard. In fact, Cornish cross hybrids do best when they're kept in a very small confined space with around-the-clock low light. If you haven't guessed yet, those are exactly the conditions that chickens are raised under in questionably ethical mass production farms. They will often fall subject to disease, a heart attack caused by stress, and broken bones. The intentional breeding to produce Cornish cross hybrids has resulted in an unstable body structure.

Do You Need a Rooster for Hens to Lay Eggs?

A common misconception is that you need a rooster on the property for hens to lay eggs. Local feed store owners often perpetuate this belief, and there's a substantial amount of misinformation about this subject online as well.

A hen does not need a rooster to lay eggs. However, if you're planning to raise chickens, a rooster is absolutely necessary. The only thing that a rooster does is fertilize the eggs. However, many homesteaders believe that having a rooster on the property serves as a type of security for the hens or that perhaps the hens are more productive when a rooster is around because they feel that the environment is safer.

When deciding if you're going to have a rooster or not ask yourself these questions:
- Will I be raising baby chicks?

- Will my hens be near potential threats? (Even perceived threats such as a dog.)
- Do I want fertilized eggs?

Silkies and Brahmas – Momma Hens

We just briefly touched on the purpose of roosters on a homestead. Now let's touch on the role of momma hens. Whether you choose to use heritage or alternative chicken breeds for your egg production, it's not likely that those hens will sit on their eggs. If you do happen to have a particular chicken that is brooding, it's not necessarily a good thing. A brooding chicken may become aggressive if you attempt to collect eggs and may become depressed or stressed if you continue to remove their eggs. You don't want a brooding heritage or alternative breed hen. So, what do you do if you want to raise your own chickens?

The answer is to bring in one or two momma hens. Momma hens are usually Silkies or Brahmas. These are chickens that you probably won't rely on for egg production or meat production. What they do is mothering work. They sit on eggs, protect them, and they are very good-natured. They're so desirable that some people choose to keep them as pets.

Silkies

A silkie is well-known for its silk-like features and docile temperament, and they love being held. Silkies don't look like your normal chicken. They are very fluffy and usually grow too big, although certain sub-breeds may be larger than standard.

Brahmas

Brahma chickens, in contrast to Silkies, are typically much larger than your average chicken. In fact, one of their giant breeds had quite a bit of news attention as the chicken went in full height to about three feet tall. On average, a Brahma chicken will grow to about thirty inches tall and weigh over twelve pounds. They also have very long lifespans, with it taking nearly three years to reach their full size.

Other Factors to Consider When Raising Chickens

Along with the types of chicken you have to choose between, you'll need to consider a few other factors. For example, if you have coyotes or wolves, you'll want to ensure that you have a very secure chicken coop. Additionally, you'll need to plan for your egg storage. It's not likely that you'll eat hundreds or even thousands of eggs in a year. Some people give them away, others sell them in local farmer's markets, while some store them by pickling or freezing.

These are just a few factors to consider, and they might help you determine what type of coop to build. For example, if you get a lot of rain, then you should seriously consider a chicken coop that is raised a few feet off the ground, so it's not flooded frequently.

Build a Chicken Coop

Every chicken coop has the same basic elements. You will create a structure to protect your chickens from bad weather and give them a safe space to lay their eggs. Your chicken coop does not need to be complex or elaborate. In fact, the simpler the chicken coop is, the easier it is to maintain. It is nice to imagine giving your chickens a lush mansion with every add-on imaginable, but that's not reasonable.

You'll want to get started by determining how many chickens you'll have. You want to give each chicken about four square feet of space to walk around. There is also the debate on the number of nesting boxes you need. Chickens will share nesting spaces, but if you plan on growing your flock later, it might be better to design in more nesting spaces earlier on than to try to add them in the future.

There aren't set names for the different styles of chicken coops, but we can lay out a few basic shapes for you to get a feeling of what options are available. The most common style of a chicken coop is a shed frame; the coop itself simply looks like an outdoor shed, with an outdoor wire space set next to it. There is also the a-frame style, which looks a bit more like a propped-up tent. You can create a chicken coop in a gazebo style which is typically all open across the sides, and then there is the off the ground "Palace" structure. The most functional and the most economical for space is the shed style structure, so that's what we'll cover here to help you learn how to build a chicken coop. If you are interested in an A-frame or gazebo style chicken coop, then you might browse around online for a pre-made structure or a set of plans. Building any structure other than your standard shed style might call for more advanced building skills. Of course, any skilled builder can create custom designed and uniquely shaped chicken coops that fit their yard.

After you decide how many chickens you will keep and the style of shed that you would like, you'll need to decide where to place it in your yard. When you're working with backyard homesteading or creating a mini-farm, location is a really big deal. Your flock has certain demands, and that might change how you arrange your backyard homestead. Your flock will need sunshine, access to shade, good airflow, and low noise levels, and you'll need easy access so you can get in and out comfortably. Try to consider a few different locations and spend some time in those areas first to get a feel for the noise level and airflow.

Next, you'll want to plan out your coop, and when you're using a basic shared structure, it is as simple as four walls, a floor, a roof, a caged outside region, nesting boxes, and a door. However, you will need windows or slats for ventilation, and a designated spot for your feeder and waterer.

The extras that you might consider adding into your chicken coop involve a perch area, dust bath, lighting, and poop boards. If you're considering adding in any extras, prioritize the lighting and the poop boards. Lighting is an element that can help improve egg production, particularly during the winter months, and what you want is to bring UVA and UVB lighting directly into the enclosed area of the coop. Having lights in the coop means you need access to electricity, and if you have harsh winters, not only for lighting but for water heaters. If at all possible, ensure that you can get electricity to your chicken coop. Now poop boards are a small extra that doesn't take much work and can save you a lot of time in maintaining your homestead. A poop board is placed below the perching area and should easily slide in and out of the chicken coop.

Once you have your plan sketched out, you'll be ready to select your materials. Most who are building their own coop prefer wood, and you don't have to worry about the type of wood so much with chicken coops. You can choose something affordable. However, there's no need to bring in wood such as Cedar that would stave off pests. The chickens will gladly welcome insect company. In addition to all wood materials, you will need chicken mesh or chicken wire. This type of wire is woven loosely enough that good airflow can get through it, but not so loosely that the chickens can escape.

The best possible size for most mini-farms is a 4'x4'x6' structure for ten chickens, and then an additional 4'x8' chicken coop run. There are countless free plan options that you can refer to when building. However, if you can set up a floor, four walls, a door, and a roof, you'll have no problem. Perhaps the best resource in this regard is the numerous DIY tutorial videos. Even with a written plan, if you're not familiar with building yet, this is a great place to get started, as it's best to watch someone build a structure or to get hands-on help.

If you're worried about getting your materials and equipment, note that you can rent most of what you need. Local hardware stores and even the larger chains will cut materials on-site and rent large tools. That means that you don't have to invest in a circular saw or heavy equipment. Chickens can play a substantial role in your homestead, and it's worth the effort to build them a good coop.

Chapter 6: Preparing Your Kitchen for a Homestead

Most people use their kitchen for general cooking and food storage. But a homestead goes far beyond general use. When you have a homestead, you must have some joy in your kitchen, because you're going to be using it for much more than general use. You won't occasionally be throwing something into the oven or turning on the crockpot the night before; instead, your kitchen will be used for other homesteading jobs besides cooking. You will use it for canning, jarring, other forms of food preservation, and even processing food from a raw state into another state, such as turning milk into yogurt.

Your kitchen will need to handle the storage of fruit, vegetables, and livestock goods with ease. That means you'll need to have all the necessary tools to process these goods, and then storage abilities to hold and manage them properly. We'll get further into long-term storage techniques in a later chapter; for now, we'll focus on what supplies you will need and how to get the most out of your kitchen's space.

When preparing your kitchen for a homestead, it should, of course, be highly functional, but it should also be a safe haven. To make sure that you can enjoy your kitchen and that you minimize any kitchen-oriented frustrations, start by making sure you have all the right supplies.

Checklist of Supplies

Part of homesteading is about learning to use what you have. You should have the basics, and then a few things that can generally help make life easier or your kitchen safer. This list has a few things that you could probably do without but that you probably won't want to go without when it comes time for cooking, organizing, and preparing food.

Before we dive right into the list, let's put a special note on quality. Buy what you can afford, but if possible, buy the best you can afford. It is far better to have a high-quality cast iron skillet and Dutch oven combination than a set of cheaper pans and a cheap Dutch oven. Also, when buying higher quality items, it typically means that you're going without optional items that could clutter your home. Then there is the element of care. When something is more expensive or higher quality, you generally care for it more. Again, cast iron is a prime example of this in action because cast-iron requires maintenance after every use, but that one pan or pot can last for over a hundred years! They last so long because the people using them know that maintaining them is vital.

So, what exactly do you need to get started? This list here lays out everything that would be nice to have in a kitchen. Typically, you'll want at least one of everything, and usually, you can make do with only one item for each purpose. For example, you don't necessarily need three sauté pans, even if they are of different sizes.

Kitchen Supplies List

- Knives – must-haves are a chef's knife, bread knife, and a cleaver. However, a full set with paring knives and steak knives is an excellent choice as well.

- Shears – kitchen shears and shears for breaking down poultry.

- Pots – A large pot for water-bath canning and making stock, but smaller pots are very useful as well.

- Wooden utensils – opt for a traditional spoon, slotted spoon, spatula, and pasta spoon. Wooden utensils are generally better for all types of cookware compared to their metal counterparts.

- Cast-iron – must have a skillet, but a Dutch oven can be extremely useful.

- Measuring cups – Any set will do; if you're baking a lot, a 1 ½ measuring cup is very useful.

- Strainers or china caps.

- Mixing bowls.

- Cutting boards.

- Kitchen scale.

- Rolling pin.

- Bread pans – most homesteaders make their own bread, and for that, bread pans are extremely useful.

Kitchen Equipment List

- Oven and range/stovetop.

- Refrigerator.

- Freezer – with homesteading, a chest freezer is exceptionally useful!

- Sink – Consider the amount of time you'll spend in your kitchen; a bigger sink can be a huge stress relief from feeling like the sink is always full.

- Mixer – Standing, mounted, or handheld options are all good options. If you want to bake a lot, consider investing in a Kitchen Aid or commercial model, it can save a substantial amount of time.

- Blender.

- Food processor – save time on all that chopping!

- Thermometer – for baking and candy making, get a candy thermometer, not just a meat thermometer.
- Scrap bucket or kitchen compost box.

Optional Kitchen Equipment – The Luxuries in a Homesteading Kitchen

- Bread maker/bread machine – most will prove and bake the bread.
- Pressure canner – do away with canning in a pot.
- Dehydrator.
- Meat grinder.
- Sausage maker.
- Pasta maker – add-on options for standing mixers.
- Meat slicer.
- Food saver (vacuum-packed sealing system) – can severely reduce freezer burn and demands for storage space.

How and if you use these tools is completely up to you. Do consider your personal preferences, but don't be afraid to step outside of your comfort zone. Many people that haven't used a mixer before may favor a hand mixer and eventually upgrade to a standing mixer. Just the same, it's likely that you've never used a pressure canner or a bread maker. Remember that part of the satisfaction of homesteading is trying new things and learning how to make your own goods rather than running to the grocery store. Self-sustainability will always call for additional learning and a bit of fun.

Homesteading Tools and Supplies for Your Kitchen

There are some tools that only homesteaders will need. They can be used for humane slaughter, storage solutions, and long-term preservation. These tools provide very specific functions that you will likely need even if you're doing the very basics of homesteading, such as gardening and keeping chickens.

Homesteading Kitchen Tools and Supplies

• Yogurt culture – comes in both dried and liquid forms.

• Electric yogurt maker – not a must-have, but certainly a nice addition to your kitchen.

• Yeast – for baking bread.

• Glass canning jars – getting a variety of sizes is important, you'll also need lids and rings.

• Sheet trays – the more, the better.

• Canisters – that will keep dry goods dry, and pests such as moths out, and reduce the likelihood of weevils.

• Supplies for cheese making – curd knife, cheese molds, cheese press, cheese draining mats, cheese wax, cheese wraps, and butter muslin.

These supplies aid in a lot of dairy handling and storage. Even if you don't have a dairy cow on your farm, you can use store-bought pasteurized milk for making yogurt, and cheese. If your goal is to one day have a dairy cow, you can certainly work on your yogurt and cheese making skills now to be prepared for when you have a substantial amount of milk in your home.

Specialty Kitchen Tools for a Homestead

Some homesteads require very specific tools for the kitchen or related to the kitchen. These tools are for more advanced homesteaders and may not directly apply to your initial homesteading plans.

For rabbit culling, a means to "remove from the herd," which can mean giving the rabbit a new home or using it for meat. Typically, when you hear homesteaders refer to culling or dispatching, it is the nicest possible way to replace the word slaughter. Many people do get attached to their animals and still use them for their intended purpose (meat production).

A specialty tool for dispatching or culling rabbits is called the Hopper Popper. This tool provides a humane way to dispatch the rabbit, and the smallest size available also works on chickens. Although this isn't particularly a "kitchen" tool, many people mount it either inside a shed or on the side of the house away from the remaining rabbits or chickens, and it is a solid alternative to less-pleasant ways to handle dispatching.

Composters are another specialty tool, and you can keep a separate kitchen compost and an outdoor compost. But, as a specialty tool, you might consider a closed-compost box which should help reduce any odors in your kitchen as it doesn't directly open to your compost within the box, but rather has a two-step drop process.

How to Get and Stay Organized

Getting organized is one thing; staying organized is another. But, with some good planning and storage solution options, you can make staying organized easy. As you go through what you'll need in terms of space, you'll be able to assess better how you can manipulate the kitchen space you have. There is absolutely no reason to do a complete kitchen renovation to start a homestead! However, you might consider adding shelves, a wire rack, and pantry storage solutions such as stacked shelving.

Step One: Assess Your Space

You'll want to carefully evaluate where and how you currently use your cabinets and any shelving in your kitchen. Pantries are prime spots for items that you use every day! Walk through your kitchen with a notepad and document what items you keep and where you keep them. That box of stuffing that will sit on your shelf for months until the next holiday rolls around certainly doesn't deserve prime pantry space. Your cupboards don't have to only hold plates and glasses.

After you have taken stock of your kitchen and your kitchen space, go through each item and determine how often you use it. If you really do reach for your chopped walnuts every morning because you love them with breakfast, then keep them out and available. But if you only use them once a week for baking, or even less frequently than that, consider moving them into an area that is less convenient to reach.

Step Two: Declutter

Clutter is nothing to be ashamed of. Everyone has it. But when you're preparing your kitchen to become a critical part of a homestead, it's time for a good cleaning out and doing away with all of your current clutter. Just keep in mind that no matter how hard you try, clutter will always resurface, so don't get discouraged if you need to declutter again in six-months.

When you're decluttering consider following these guides:

● Keep only one of the things you need (one 1-cup measuring cup, or one slotted spoon).

● Stack inside and stack up (stack your cups in your cupboard, stack your pots together too)

● Ask, "Do I have something else that does the same thing?"

o Forget the hot debate between having a punch bowl and a serving pitcher, or between having a rice cooker and a pot. Some things don't need a fancy device or a duplicate. Now is the time to eliminate the unnecessary (especially single-function) kitchen gadgets.

● Ask, "How often do I use this?"

o Make piles for reorganizing your kitchen based on how frequently you use items. Daily items deserve locations as close to the counters or as easily accessible as possible. Weekly items deserve the outlying space, and monthly items need to go somewhere hidden or tucked out of sight.

● Ask, "Where is the most logical spot for this item?"

o Make sure that your most-used utensils are where you need them the most, such as near your stove. Then consider keeping others like things together where they best serve their purpose.

The decluttering and reorganizing step isn't about throwing everything out. It is about making sure that what you have is either handy or logically stored.

Step Three: Organize Everything – Including Your Food Storage

Before you put away everything that you dug out during the decluttering step, mark your cabinets, cupboards, and pantry with what you would like to put there. It will help you avoid derailing and just putting things back where they always were. Don't be afraid to keep things on the counter if you really do use them every day. If you make rice every day with dinner, then leave your canister of rice easily accessible on the counter!

Tips for Organizing

• Store dry food in large clear containers or canisters so you can easily see how much you have.

• If you buy bulk, then keep a "kitchen container" in your kitchen and store the rest in a bulk storage area.

• Keep your like items together – cans go with cans, grains go with grains, beans go with beans, and so forth.

• Keep a printed inventory list on a clipboard hanging somewhere easy to see while you are in the kitchen. Then, when you've used all but the last of something, you can add it to your shopping list, make more, or re-plan your planting, so you don't drop in quantity so quickly.

• Don't be afraid to store things in different places; for example, if you use honey every day, but buy it in bulk, then keep your bulk supply tucked away and enough to last you through the week out where you can access it easily.

• Add wall shelves – you immediately add storage space, and with Mason jars or canisters, the storage solution can look pretty snazzy too.

- Add stacked storage in your pantry - with pantry stacking shelves or stacking drawers, you can nearly double your pantry space.

When working within your kitchen, you might need to reorganize a few times before you find a complete system that works for you. At this time, if you're not sure how often you'll use something or need to access it, use sticky notes. Put sticky notes on items and then remove them when you have used them. At the end of the week, anything without a sticky note deserves a prime spot. Then go through and do it again for the month with a different color of sticky notes on the already approved items, so you're not confusing your most used items with your frequently used items. You can do this to help you manage your storage rotation as well.

Ultimately, your kitchen will become the indoor hub of your homestead. Don't be afraid to make changes to your system if they're going to help you better track and monitor your inventory and easily access all of your kitchenware.

Chapter 7: Harvesting and Preserving the Fruits of Your Labor

Even with excellent planning skills, there's no way that you can eat your way through everything that comes out of your garden or coop daily. When you factor in that your chickens will probably lay one egg each per day, and then you'll probably have between four and six chickens, you'll end up eating four to six eggs a day. That's a quick recipe for high cholesterol and a boring diet. However, using modern and traditional methods of food preservation, you can ensure that your pantry and refrigerator are stocked with your favorites year-round.

Home food preservation techniques are tried and tested. Hundreds of generations have gone through with some of these techniques in the past and lived. Although they may seem new to you, or even completely foreign, give them a chance. You might learn that something that seemed scary is actually something you're good at. Canning, for example, scares a lot of people because of the risk of botulism and other illnesses arising from improper canning. But canning is also very easy, and you'll probably do it right on your first try. So, give these options a chance and take advantage of every opportunity to keep your food for as long as possible.

To get started, here is a list of different ways to store and preserve food:

- Canning – using a combination of pressure and natural interaction between the materials.
- Freezing.
- Dehydrating – simply running warm to hot air around fruit, vegetables, and meat to produce dried versions.
- Root cellar – originally an underground structure, although modern renditions have used sheds and above-ground structures, providing a cool and semi-moist condition similar to your refrigerator vegetable drawer.
- Pickling – preserving in vinegar or brine.
- Jams and jellies – When unopened, homemade jam can last between one and two years on a shelf.
- Salting meats – meat can last between 1 and 2 months when salted, even without refrigeration. Salting meats is an old-age method of food storage.
- Fermenting – the process of using yeast or grain to cultivate beneficial bacteria and prolong the life of food. Common examples of fermented food include sauerkraut and kimchi. Fermented foods stored in a cool dark location can last between 4 and 18 months.
- Water kefir – probiotic culture lasting about 2-3 weeks in the refrigerator and up to two months in the freezer.
- Milk kefir – fermented milk consistent with drinkable yogurt lasts about 2-3 weeks in the fridge or up to 2 months in the freezer.
- Jerky – dehydrated meat, usually good for 1-2 weeks
- Sausage and hamburger – easier to freeze and store in reasonably sized portions.

If any of these sticks out as a reasonable option, then it's time to put it into your homesteading plan. Below is an in-depth explanation of all of the most popular options for preserving and storing food for a homestead.

Canning

Canning food involves applying heat to the food while it's sealed within a jar, destroying microorganisms that would otherwise lead to food spoilage. You want to use a proper canning technique, and two well-used techniques to accomplish proper canning are water-bath canning and pressure canning. Other techniques are not reliable and may put the safety of the food at risk. With canning, in any case, it is important to not fully close the rings around the lids as it can result in the can appearing to be sealed when, in fact, it is not. Close your rings tightly, and then give them a gentle nudge to loosen them just a bit.

Water-Bath Canning

Learning about canning is where most homesteaders begin because all you need is a large pot, a lid, and a rack. For effective water-bath canning, you'll place your jars with the lids on in a pot of boiling water until the inside of the jars reach 212-degrees. The time that the jars must remain at that temperature varies based on what you're canning. This method works exceptionally well for fruit, pickles, tomatoes, and other very acidic foods.

Pressure Canning

With pressure canning, you need a pressure canner as a piece of equipment. The pressure canner creates a locked compartment filled with steam and pushes the internal temperature in the jars to 240 degrees. At the same time, it applies a specific pressure that is based on the weight within the device. On these devices, you'll see a gauge that indicates the weight. This method is best for vegetables, poultry, fish, and other meats.

Pickling

There is the traditional form of pickling, but because of recent developments, there is also quick pickling. You can pickle nearly any type of vegetable, and you can pickle boiled eggs. The process of pickling is generally straightforward. Place the cut vegetables in jars of vinegar or a vinegar mixture of water, vinegar, salt, possibly sugar, and a wide variety of spices. Most people use pickling spice no matter what, but you can also include garlic, peppers, dill, and more. Then, store them in the refrigerator.

If you're more worried about keeping your fridge space open, then you will need to pressure can them or give them a water-bath canning process to ensure that they're properly sealed.

Freezing Foods

Freezing food is about much more than throwing something into a "freezer-safe" bag. You'll want to use freezer paper; wrap meat in freezer paper and butcher's paper, and then seal it properly. And you'll want to ensure that you're reducing as much air exposure as possible. When you freeze anything, you want to reduce the air exposure, because it is very cold, dry air that "burns" your food and leads to quick decay while in the freezer. Also, be sure that you never over-stuff your freezer because it needs airflow to slow any damage to the fresh food.

Tips for Freezing

- Freeze food at the peak of freshness.
- Always wrap food tightly and in multiple layers before putting it into the freezer.
- Keep your freezer full, but not stuffed.
- Boil and blanch your vegetables before freezing.
- If possible, store in vacuum-sealed bags.

When freezing vegetables, it may feel counterintuitive to cook them first. But you only want to boil your vegetables for a maximum of three minutes and then immediately submerge them in an ice bath (blanching). This takes the vegetables to their ideal state, and then when they freeze, they should retain that state.

Freezing berries or fruit is a different situation; you can typically freeze fruit without any cooking or blanching. In fact, cooking immediately begins to break down fruits as they're much more sensitive than most vegetables, especially root vegetables. You may need to approach each fruit slightly differently, though, because you don't want just to throw them into a bag. Ideally, you will lay fruit out on sheets, allow them to freeze through completely, and then put them into the freezer with proper layers of protection to avoid freezer burn.

Meat can do very well in the freezer, but again, the important aspect is layers. Freezer paper can make a substantial difference in quality. Then there needs to be another layer, typically of plastic wrap or wax paper, and then finally, storing them in a freezer-safe enclosure such as a freezer-safe bag or container.

Root Cellars – What You Need to Know

There are different types of root cellars, and not all are made equal. The particular problem that people face today is zoning and permits. Because root cellars are traditionally underground and should be tall enough for you to at least crouch standing up, you may have a rough time avoiding lines, plumbing, and more in your yard. However, if you find a local contractor, they may be able to help you determine if you are okay with creating a root cellar on your property. Root cellars are well worth the investment as they can help keep food fresh and at its most natural state for an extended period for time.

Root cellars are what people used before refrigerators were widely used. Almost everything can be saved in a root cellar. You can store canned goods, jarred goods, medicine, and produce such as carrots, turnips, potatoes, pumpkins, squash, and more. The absolute best environment for a root cellar is between 32 and 40 degrees (F) with 85-95% humidity. Because the humidity levels are so high, there is a minimal loss of moisture from the food. The low temperatures keep the decomposition rate at a minimum. The result is that the foods stored in a root cellar release ethylene gas at a much slower rate than if they were in any other environment. That gas then escapes through the ventilation.

Now, you must keep fruit and vegetables separate because of the drastic variances in decay time. They each release ethylene at different rates, and storing them together can cause both the fruit and vegetables to deteriorate more quickly.

Extending the Life of Dairy

Often, dairy is short-lived, with most store-bought milk having a shelf-life of between five and seven days, whereas yogurt is usually stable for one or two weeks, and cheese can last in a freezer for between 6 and 8 months. If you have a milking cow on your homestead, it is well worth the effort to learn how to make cheese and yogurt to get the most out of your milk production.

It is important to mention here that pasteurizing milk does not make it last longer. Pasteurized milk will typically have a shelf life of a week or less, but organic milk treated in other methods can last longer. A method called UHT or Ultra-High Temperature pasteurization allows the milk to remain shelf-stable, when properly refrigerated, for up to six months. UHT calls for super-heating the milk to 280 degrees (F) for two to four seconds and then rapidly cooling it back to 39 degrees, which kills any bacteria.

Unfortunately, at this time, there is no fool-proof way to accomplish this at home. Some use pressure cookers, and others use "Instant Pot" pressure cookers, but the result is the same, the milk tastes cooked because it is so difficult to chill it at home rapidly. It is generally recommended to avoid trying to UHT at home. However, there is hope among the homesteading community that there will soon be more options available for those who want to UHT at home. Maybe a new gadget or device that will assist in both the heating and the cooling, but as of 2020, the technology just isn't available on a residential scale.

Home Food Preservation Safety

Always ensure that you're focusing on safety. Canned foods come with the risk of botulism and other illnesses, as well as inherent bacterial growth and fungus. Pickling, when not done correctly, can lead to sour vegetables and wasted shelf space. Every method of food preservation comes with risk, even when you simply put things in the refrigerator.

When opening any food product that was stored for a while, smell it and explore it visually for signs of quality deterioration or decay. Ideally, these methods of preservation will lock your products in time at the peak of their freshness or turn them into another food product, such as a fresh peach to jarred peach slices. These preservation methods can help you keep some of your crops in stock all year. Peaches make a great example because they only ripen through the summer, but with freezing, jarring, and making jams, you can use up your delicious peach stock all year.

Explore the different preservation options and see what you can manage in your kitchen and with your dedicated storage areas. There are more options available, too, if you're looking to break past the beginning level of homesteading and into the more advanced methods of food storage and preservation.

Chapter 8: Making a Profit from Your Homestead

Most people don't start a homestead with the dream of making a profit from their land. In fact, most people start out just hoping to curb some of their grocery bills and live a more self-sustaining lifestyle. But you can make the dream of living off of your homestead a reality. All of these methods of making a profit from your homestead do require work and dedication. However, they can also be very feasible ways of making money from something you're already doing.

If you were still in the planning phase of creating your homestead, then you might create a type of business plan to assess different ways that making a profit might fit into your homestead designs. Even people who are working on only a half or quarter of an acre can find many ways to cultivate a profit and a beautiful garden. Keep in mind that every homestead will face unique limitations, and your community may play a part in what works and what does not work for you.

Branding Your Homestead

If you're going to have any products that could generate revenue, you need a brand. A brand provides people with a way of associating with your homestead and products. There are countless examples of outstanding brands out there in the world, but your brand must represent your lifestyle.

When creating a brand, consider what values drove you to eventually commit to growing your own homestead. Sharing your story about how you became a homesteader and how you developed into a business could be very important for your customers. Generally, customers, even those who are interested in organic or homegrown products, want to understand the business behind the product. Giving your future customers a bit of insight into your life and your values can have a dramatic impact on your business.

What you want to achieve is what marketers call organic branding, a brand that develops as a result of the meaning and purpose of the product. It means that the brand has a unique voice, an attitude, and a set of views on ethics and cultural elements within the community. As the person selling goods, you're the one who will determine all of these things.

Branding does not mean that you need to have customized stickers, business cards, packaging, or cheesy commercials. While it certainly is nice to be able to sell T-shirts or to slap on a sticker with your homestead logo or a fun name you've given to your homestead, it isn't necessary. You simply need a name, an image, and an attitude to attach to a stall at a local farmer's market or an online craft shop. No matter what the platform or what products you're selling, you'll need some branding.

Selling Your Goods

Selling goods is the most direct way to make money from your homestead. It's a great way to get rid of excess inventory, and it can help reduce the time that your harvest sits on your shelves. If you know that you're going to have a plentiful amount of harvest every season, then there's no reason not to sell what you don't need for your household. The great news is that you can sell almost anything that you grow at home. As long as what you're selling is legal in your area, then it may be as simple as finding a local farmers' market and completing an application with your city or County officials. However, there are areas and there are items that do have legal restrictions.

We'll get into more details about legal restrictions a little bit further into this chapter, but to give a quick example, raw milk is one of the things that are illegal to sell or distribute on a national level. U.S. Federal guidelines given by U.S. Food and Drug administration ban the sale of raw milk, and the solution to selling milk would be to pasteurize it. Many legal restrictions on selling food or goods from a homestead have a clear-cut solution, so don't get discouraged if you see that there might be a few obstacles in your way.

What to Sell

So, what can you sell? You can sell almost anything, but here is a long list of things that are available on almost every homestead.

- Sell extra produce
- Sell seedlings
- Sell compost and compost materials
- Sell manure (bunny droppings, cow droppings)
- Sell your flowers
- Sell farm-fresh eggs
- Sell baby chicks
- Bottle and sell goat milk
- Bottle and sell cow milk
- Sell homemade butter
- Sell jerky
- Sell fresh herbs
- Sell dried herbs in premade spice mixtures
- Sell large poultry such as geese or ducks for hunting
- Sell potted plants

- Make and sell specialty pasta or pesto

Selling what you produce on your homestead is one thing, but you can also use it as another selling platform. If you have space for a pumpkin patch, it is very reasonable to open up a side gate to your pumpkin patch and charge neighbors to pick their own. Or you can do the same thing year-round with a U-pick farm where people throughout your community can enter through a side gate and then pick vegetables at their leisure.

There are some products that you can sell that would be considered special to that particular homestead. Here are some examples of homestead specialties that don't fit the mold for every homestead but can certainly be accessible products from your homestead.

- Honey from your bees
- Beeswax products
- Raise animals for slaughter for other people (similar to leasing out a pasture)
- Fishing on-site (if you have a fish farm)
- Fish (if you have a fish farm)
- Fishing bait (cultivated insect-life)
- Raising working dogs (more on this in Chapter 12)
- Building vertical gardens

All of these items refer to add-ons to a basic homestead. Not every homesteader raises bees, but those who do can benefit greatly from their honey and beeswax products. Additionally, not everyone is involved in aquaponics, but if you have a fishpond, then you have access to a wider variety of products to use to make money off of your homestead.

Raising animals, in particular, is another great way to make money from a homestead without technically selling a product. What you're selling at this point is a lease on pasture or herd-sharing, and the time spent raising and caring for the animals. Many people would love to have a home-raised cow or pig for fresh organic beef and pork at home, but don't have space or can't manage the maintenance for the animal. As a homesteader, you're equipped to do just that. There's also the option of raising working animals. For example, working dogs don't produce meat, but they are exceptionally helpful on farms and homesteads, and not every homesteader has the patience to raise their own working dogs.

Crafting and Creating

It's probably a safe bet to say that if you can build and sustain a homestead, you might be pretty crafty too. Crafting goods is one of the really fun elements of homesteading where you take raw goods or your excess materials and make something completely yours. Crafting is also built into the self-sustaining element of a homestead, where you may need to, for example, build a chicken coop instead of purchase one. When you turn to craft and creating, you're doing a lot more than selling raw or nearly raw goods.

Sell these to make an extra bit of income from your homestead:

● Sell jams, jellies, or preserves.
● Sell candles.
● Sell Chapstick or other beeswax items.
● Make your own luscious hair and skin products from all-natural ingredients.
● Sell homemade soaps.
● Make fishing lures out of chicken feathers.
● Repurpose wood into different pieces of furniture or home décor.
● Sell crochet or knitting projects.
● Sell quilts and other simple-living luxuries.

This list is not all-inclusive, and you can add nearly anything that you can think of to it if you can make it on your homestead or in your kitchen. The idea, of course, is to use goods that you've created on your homestead to generate profit with a bit of creative flair. For example, homemade hair and skin products have a big market on websites such as Etsy because people want alternatives to the chemical-filled products you find on the shelves in stores. Of course, nearly everything on this list entails a skill you may need to learn. Skills like candle or soap making have been all but lost in the last couple of centuries. However, once you learn how to make them, you can do it again and again and continue to generate profit from that same skill set.

Craft and Specialty Fairs

All of the platforms listed in the section below are still options for when it comes to selling homemade goods. Farmer's markets allow for homemade products as well as raw goods. But there is a unique sales platform available for homemade products that don't usually apply to raw produce and raw goods. Craft and specialty fairs usually come around at specific times of the year, and it's the opportunity to set up a booth and sell your goods over the course of a few days. Specialty fairs, such as renaissance fairs, are additional opportunities. If you can make fishing lures or quilts, then you have a place in most specialty fairs that might come to your county.

Understanding Cottage Food Laws

Although initially it seems that cottage food laws are restrictive and troublesome, they are much more lenient than regulations that small business owners face. Cottage food laws allow people to sell homemade food to the public without going through extensive licensing. The primary focus is to understand the federal requirements and know which foods require labeling and which foods are outright banned. These laws vary from state to state.

But mostly, cottage food laws give an outline on which food needs to be refrigerated, how to process foods by canning, and other elements regarding spoilage. Some foods are not allowed to be produced in a home kitchen because they offer very specific health risks. However, with some permits or certain licensing, you may be able to sell these at farmer's markets or through similar platforms. Always be sure to research the cottage food laws in your state thoroughly.

Teaching and Experience Sharing

Chapter 11 will cover how to share your learning experiences, and how you can make your experience-sharing profitable. For now, we will go through the basics of teaching and experience-sharing as something you can make a profit from, and the many different ways to do it. You have the option of starting a blog or giving classes or even getting involved with local neighbors who want to learn how to homestead. The real element to consider here is that you shouldn't hesitate to charge for your time; even though you're not selling an outright product, you're selling a skill set.

Helping Others Learn Through Courses or Lessons

When just starting out, you should sit down and carefully set out what each lesson should accomplish. After you determine what each lesson will accomplish, you'll need to determine how you will communicate the information in each lesson and what activities or exercises you'll provide to help build that particular skill or bit of knowledge. What this is doing is writing a course plan and multiple lesson plans. You can find many different lesson plans online which can serve as examples.

If you do decide to give courses or lessons on homesteading, you need to think about the learning environment carefully. For example, if you're giving courses and charging money, your home kitchen may not be suitable; you may need to use a community kitchen or a commercial-grade kitchen because, in some counties, there are strict health-code requirements. When working or giving classes on your property, you may take greater steps to ensure that the grounds are safe and there is minimal risk of injury. Additionally, you may need to temporarily rent a commercial or commissary kitchen where you can host classes for kitchen activities in a food-safe environment.

Where to Sell

Getting into where exactly you can sell your goods can be a bit complex as local governments have varying laws and restrictions. By and large, you should expect to be able to sell the majority of your goods at a farmer's market. It may be a good idea to branch out and attend multiple farmer's markets, be they local or in more distant cities. However, if you look beyond farmers' markets, you'll have options to sell online, and you may open a small shop if it seems profitable, or you may sell out of your own home if you can manage it in your area.

Farmer's Markets – These markets are specifically for locally grown and produced items. Farmer's markets are legal, and on a federal level, the government sees them as a way to promote regional agriculture and assure a supply of fresh and local produce for residents. Typically, farmer's markets are restricted to small family farmers and homesteaders as they don't have the opportunities to sell in larger grocery chains. When setting up with the farmer's market, there are restrictions. You can't have a middleman, or a broker, involved; if a city tells you that it's required, then they're not conforming to federal law. There is the chance that a County agent or federal agent may visit your land to ensure that what you're selling at the market was grown or produced by you. Typically, farmer's markets will have a farmers' market committee that will have a combination of market growers and consumers that enforce rules and policies specific to that association.

Personal Online Shop – It can be difficult to get started, but in the long-run, it is usually easier to sell goods online than it is to run a small shop. A personal online shop will require a dedicated website and the use of an e-Commerce facilitator such as Shopify or Magento. Although it can seem complicated at first, with a consultant or a knowledgeable friend you can set up a website for less than a few hundred dollars in just a couple of hours. Websites are not as expensive as they used to be, and they're not as complicated to set up as most people believe. Once your online shop is set up, it is as easy as doing some simple maintenance to ensure that the products you have available are also available online for customers to view. You do need to be very careful with what you post online for sale, however, as there may be restrictions on cottage food laws and how you can ship certain items safely.

Online Crafting and Homemade Goods Platforms – Platforms such as Etsy, Aftcra, Artfire, Cratejoy, and Absolute Arts allow people to sell handmade products online. Using online crafting or homemade goods platforms can be a much easier way to get a jump start on selling goods online without having your own dedicated website. These platforms do take a small percentage of what you earn and may have additional fees based on shipping and other services. Explore which platforms you would be interested in using and research their fees and user agreements carefully. For example, Etsy allows its sellers to sell food and edible items as long as they follow government regulations. Their Seller Handbook offers specific information on how to sell, ship, and package food items in ways that comply with federal government regulations.

Selling Out of Your Home – For home-based businesses, there's a handful of unique struggles. First, every county can set their own restrictions. Second, each county and state can have different permit requirements and, finally, you may be required to have certain insurance policies, such as liability insurance. When selling food or homemade products out of your home, you need to register as a business. Additionally, you'll need to go through your city, county, and state to determine if there are permits necessary for you to sell from your residence. For example in California, one of the most restrictive states, you will need to have one type of permit to sell directly from your home and at farmer's markets, and another permit if you plan on selling through those platforms and additional venues such as a local store or restaurant. And you'll be asked to provide information for all of your ingredients, your recipes, and where you're sourcing your ingredients, as well as having to make labels for each product you sell. This is why many homesteaders sell products directly from their front door. Your city or county may have a wide variety of exceptions available to you if you are only selling raw goods. For example, again within California, producers of raw unprocessed foods may only need a certified producer certificate rather than a handful of permits and regular check-ins from the public health inspector.

How to Build A Profitable Homestead

Whether you're still planning your homestead, or you've had a successful one for a few years, you can start with a basic checklist to make your homestead more profitable.

- What income streams can you access from your homestead (or homestead plans) without changing anything?
- What resources are available?
- Will you need additional resources?
- Define your customer (who do you ideally want to sell to?)
- Identify customer needs (e.g., they want organic options, they are trying to get away from big stores, etc.)
- How is your product different?
- How will you set your prices?
- Will you need an online presence or a community presence, or both?
- Write your business plan.

This checklist will help guide you through the process of establishing your options and eventually creating a homestead with opportunities for profit. It is important to make a business plan, even if you only continue forward as a cottage food operator, as that document will serve to help you make decisions and set goals.

On a final note, it is important to ensure that the purpose of self-sustainability within your homestead remains the top priority. If you're selling so much product that your homestead itself is withering, then you don't have a sustainable model, and you may need to scale back on your customer base or scale up your operations and become a larger business. Many homesteads are profitable in ways other than simply living off their properties. While you're still in the planning stages of putting your homestead together, it's a great time to explore how you might bring together homesteading, your hobbies, and the opportunity to bring more income into your household.

Chapter 9: 8 Resources to Consider

How and where can you start? Certainly, it's difficult to gather up the money necessary to purchase equipment, seeds, chickens, and the materials necessary to build a coop and garden boxes. But you can do it. There are many resources available to help you get started; some are financially based, while others help you build experience and skills.

There are four key elements you want to cover when you're looking for high-quality resources. First, traveling and labor assistance. Although you're not moving heavy equipment, you might be creating a lot of garden boxes, building a chicken coop, or planting trees; these aren't tasks that one person can happily do on their own. The more help you have, the more likely it is that you can develop diverse skills. The laborers that you'll find through volunteer systems may have no experience, or this may be their retired hobby with decades of experience. Always accept help happily.

Second, you need support from your county or possibly your city. Some cities, but nearly all counties, have restrictions on what you can do in your own backyard. It might seem outrageous that any government office within U.S. could say what you can or can't do on your own land. However, there are some very reasonable explanations for why you can't dig in certain areas, build in others, or even have chickens. For example, chickens are noisy, and if your city has a restriction on noise within your neighborhoods because of other residents, it may mean that chickens aren't an available option. You may not be able to dig lines for waterways or water systems because of buried gas or power lines. These obstacles may or may not be present within your neighborhood. You must communicate with your city or county and acquire any necessary permits.

Third, you need to find exemptions for farms, property tax credits, and other tax elements. If you profit or earn some type of income from your land, then you may need to calculate your income and pay taxes on that revenue. And if you give back to the community, you may have access to some tax breaks or write-offs. If your homestead becomes your business, you may have the ability to write off certain pieces of equipment. It's best to find the resources you need to figure out what may or may not be available to you. When it comes to taxes and money management, nothing compares to a Certified Public Accountant. Although you may have the skillset and resources to handle your finances on your own, always keep an open mind toward hiring a CPA if you grow or start earning significant income from your homestead.

Finally, there's the 'where to start' resources. Although this guide is very comprehensive, it might be helpful to have a few local resources available when you need a bit more guidance. Besides getting local help in getting started, you can find the support you need through these trying times within the homesteading community. These resources are more for emotional support, for someone to step in and say, "Yes, you can do it."

Let's dive into the resources and keep in mind that some are very up to date with information on modern techniques and implementing modern systems, while other resources here only refer to traditional methods or support "traditional" homesteaders.

It's vital that you cast a wide net, as it can take a while to find the right resources for your mini-farm.

Finding Help with Labor and Developing a Skill Set

The Worldwide Opportunities on Organic Farms or WWOOF, helps new farmers connect with local volunteers. These volunteers can greatly vary in experience, but often you'll exchange education and culture while working with this group. You may find that many people within your community are happy to jump in on a project and literally roll up their sleeves.

Volunteers with WWOOF will usually live with the host for a short time, and it allows them to experience life as a farmer. For you, that may mean visitors in your home that genuinely want to connect with your land and support a change in how we view and cultivate our food.

So, you would sign up with WWOOF as a host, and then essentially place an ad that will be put up onto the site where volunteers could see it. They respond through the site, and then WWOOF contacts you. It's not like Craigslist, where just anyone can contact you or view your information. WWOOF likes to urge people to give it a try and ensures that the process is as safe as possible. All volunteers go through an extensive application process, and not everyone enters the program. Additionally, most complaints will result in an immediate suspension, if not a permanent termination, of their membership. You can always do thorough interviews before accepting a WWOOF candidate.

Another way to find like-minded people that support local farms is through your local 4-H chapter or FFA groups. Although they are often much younger, you can collaborate with the group leaders to contribute to your community and future farmers or agricultural enthusiasts while supporting your local chapter.

For help through your immediate community, you might also explore your local community center. Often these centers can help like-minded adults come together. Keep in mind that if you get involved in these types of communities, they often expect you to give back. For example, you may plan part of a 4-H project with your mini-farm and continue working as an adult volunteer within the organization.

Assistance from U.S. Department of Agriculture and NIFA

The Cooperative Extension Office offers help with local and individual assistance for everything involving research, practical application, and more. The National Institute of Food and Agriculture has long led the way for researching agriculture in America. That means that they look at different methods of development, land cultivation, and other things that lead to a plentiful harvest. While working with the United States Department of Agriculture, these two authorities created the Cooperative Extension System. The System serves to translate the research of the National Institute of Food and Agriculture, with practical applications that farmers and ranchers on all scales can put to use.

You can access information and find out what monetary or informational support is available in your area by contacting the CES (Cooperative Extension System) office in your county. They provide not only access to grants and funding awards, but also education. Although there are no formal education credits that you can earn through CES, you can obtain valuable information from these groups. They have operated for over 100 years and help families, communities, farmers, and ranchers build sustainable systems within their homesteads, community gardens, and more.

Grants and Funding

When you look at the full costs of building a homestead, even in your backyard, you're looking at a few thousand dollars. The average startup costs over the first two years are about $5,000. Most of those costs happen in the first few months as you build a chicken coop, purchase seeds, obtain storing materials or equipment, and buy chickens. Now, those costs are on top of all your current bills, because you're not saving on groceries or products you can cultivate at home yet. Most people, even those living paycheck to paycheck, can make $5,000 happen over the course of two years. But there are many avenues to help you obtain funding for the initial investments. Grants and other forms of funding are usually available to help cover or entirely cover startup costs that otherwise might dissuade you from starting your homestead.

Grants.gov offers support in small farm grants, particularly if you're looking to buy land and equipment. The USDA, or United States Department of Agriculture, also provides specialized grants and small loans. For example, if taking out a loan to get your mini farm started now is part of your plan, then you need to look to find the best possible loan repayment options. An FSA loan would allow you to use the loan for buying land, equipment, livestock, supplies, seeds, and feed. There are also rural development loan and grant assistance programs that can help with things like setting up community facilities and utilities. Finally, there is the opportunity to enter into the United States Department of Agriculture Farm Service Agency for Beginning Farmers and Ranchers. This program contributes a substantial portion of support for guaranteed farm ownership through operating loan funds directly to beginning farmers and ranchers. There aren't restrictions based on how much property you have available, so you can apply while still using a backyard for your homestead.

Sustainable Agriculture Research and Education provides lists of grants that usually stem from community relationships or educational institutions.

Resources for Information on Sustainable Agriculture

When it comes to finding information on sustainable agriculture, it's easy to get lost in the sea of information, or misinformation. While this book provides a guide, we can't possibly cover every possible stumbling block through your first few years in backyard homesteading. However, we can help you find resources that have brought together years and years of research and studies so you can access reliable information quickly. The USDA Alternative Farming Systems Information Center is among the leaders in providing information on sustainable food systems in agriculture. They're recognized as the primary authority on all things related to sustainable agriculture from aquaculture to community farming support. They're a great resource, but it can be a bit difficult to navigate their website. If you're looking to use their information often, you might index or bookmark your most frequented pages on your phone or in your Internet browser.

Beginningfarmers.org is another primary resource for, you guessed it, beginning farmers. The aptly-named website is up to date with the most recent developments in small farming and agriculture. It's a great place to find news updates and get answers to questions for things such as emergency relief, or where to find a beginners training course for your area. They also have a forum for helping people find farm jobs and internships. If you're willing to pay for labor, it's a great place to post an open position and find people genuinely interested in farming and agriculture.

The USDA website has a lot of tutorials for small-scale farming. It's a great place to start on reliable DIY videos that you can follow easily. With the USDA video library, with everything from the National Agricultural Library, you can find tons of videos on remaining organic and other elements that involve expanding your mini-farm, such as aquaponics.

When you're working with agriculture, it's important to know which resources are credible and fit your style of farming. With these resources, you should have no problem finding any information you need. You can return to any of these resources again and again. What's important is that you understand that there are so many resources available. All too often, people make homesteading feel like an isolated existence. But clearly, there are many communities and many sectors of the government that offer support to individuals. As you start homesteading, keep in mind that you're not alone. Know that if you need help, you can get help, and you can probably get it pretty easily. Keep tabs for these resources as you plan out how you will start your homestead. Don't be afraid to apply for grants or funding simply because you're homesteading in your backyard. Also, don't be afraid to watch video tutorials or read the information that is more relative to large homesteads but is something that you might be able to scale down.

Chapter 10: Care and Maintenance

Care and ongoing maintenance are mandatory. You must always be thinking about and preparing for the next season. That means tending to gardens, rotating livestock, and maintaining your structures. You must consider your climate and weather conditions with high regard.

Homesteading is a process that is always on, and that means that regardless of the weather or even if you're sick, certain things have to get done to maintain your current level of success. As you're still in the planning phase of homesteading, you can take greater care now to develop a low maintenance homestead. When you have fewer demands on maintenance, you can devote your time to more productive tasks.

Caring for Gardens

When it comes to garden maintenance, you'll need to worry about weeding, mulching, and fertilizing. Weeding is very straightforward. If you see anything growing in your garden that is unwanted, pull it up. But the best way to ensure low-maintenance weeding is by properly mulching. Mulching is using a material to cover the soil to prevent weed growth and slow down water loss. Organic options include straw, grass clippings, leaves, and newspapers. However, many people choose to use bark or wood chips that have a chemical treatment.

Fertilizing certainly contributes to overall plant health, but as a maintenance measure, it promotes good soil health. Most plants, when properly cared for, won't last more than two years. While it is important to take care of the plant, it is also important to care for the soil. For fertilizing, you can use organic options such as liquid seaweed, alfalfa meal, hummus, and compost. For non-organic options, you can use any fertilizer available at local home improvement stores.

When it comes to the winter months, you'll want to ensure that you have protection for your plants. If you live in an area where the ground freezes through, you might consider using a heating wire. Additionally, if you live in an area where there are harsh summers, you should consider using shades or awnings to protect your plants during the harshest days.

Upkeep on Structures and Equipment

The maintenance of your structures and equipment is pretty straightforward. You want to create a calendar that will help you identify minor repairs before they become major problems. Any structure on your property with a roof should be inspected every six months, including the roof on your chicken coop or goat pens. It is much easier to replace a shingle than to replace the roof on anything. Additionally, every three to six months, you should walk around each structure on your property and evaluate the sides and the foundation for possible damage. You can schedule repairs during seasons where repairs may be cheaper. For example, many contractors are willing to do minor foundational repairs or wall repairs for outdoor structures in the spring or the fall at cheaper rates because there aren't harsh working conditions.

When it comes to your equipment, you'll want to work with a smaller time frame. If you have a tractor, you'll want to have it serviced every three months and use that service schedule to evaluate it for needed repairs as well. If you have additional specialty equipment, such as a wind generator or solar panels, then you'll need to explore the specific demands of upkeep for that equipment. Solar panel upkeep, for example, varies by brand, and you may need to clean them daily or use a once-a-month chemical treatment to keep them clean.

Maintenance for Livestock

Although you might feel bad when you kill a plant, livestock is a little more concerning. You want to ensure that all of your livestock has a clean living environment, but you also want to make that environment easy to maintain. You can create a low maintenance livestock plan, and doing that should naturally result in a cleaner living environment and better living conditions for your animals. It is exceptionally difficult with livestock because if one gets sick, it is common that all the rest in the flock or the herd will become sick as well. Through careful planning and maintenance, you can avoid losing an entire flock or losing an animal due to poor conditions.

Chickens

Generally, chickens are very easy to take care of. Although some do like human contact, most aren't bothered if you can't spend very much time with them. Additionally, with a feeder and access to water, they don't need very much at all. You do need to collect their eggs almost daily, but that is more for food safety. So, what should you plan for when it comes to chicken maintenance?

To create a low maintenance chicken coop, you can make the demand for cleaning much easier to manage. Chickens create a lot of waste. It can be a hassle to go through and clear out the chicken waste once a week when in that one week, it can almost cement to the floor of the chicken coop. Instead, use a linoleum floor that you can spray off with a power spray nozzle, and then add chicken litter on top of it.

There is a technique called deep litter method, which is used to create compost beneath a tall layer of litter. What happens is that you allow anything left on the floor to compost, and you add litter on top, and this makes for one really big cleaning job every few months. It really comes down to your preference in your time management, whether you would be better off cleaning it out once a week with linoleum floor or doing it once every few months with the deep litter method.

Another element of maintenance for chickens to consider is stress. These birds are very prone to stress, and when they are stressed, they stop laying eggs. One of the biggest factors that can lead to stress and molting is a poor diet. The chickens are omnivores, and they require a lot of vitamins and protein, so chicken feed alone won't be enough. Instead, you can provide them with crushed eggshells, oyster shells, and ensure that they have a chicken run where they can forage for insects. It is much easier to integrate these elements into part of the regular diet than trying to correct sickness or molting from stress every time malnutrition becomes an issue.

When it comes to maintenance for chickens, you do need to plan for what will happen when your hens stop laying eggs. When hens stop laying eggs because of their age, you'll need to choose between allowing them to live out their days in retirement in the coop or throughout the yard or to process them. It seems worth mentioning that older hens don't make for very good meat production and processing them doesn't usually yield much meat that's worth using.

Finally, you'll need to plan for extreme weather conditions, and that includes hot and cold weather. Because your chickens have a coop, they have an indoor area to go to, use that indoor area to build in extremely cold weather measures. Emergency preparedness can include heat lamps that you can plug in or switch on only when necessary. When it comes to hot temperatures, you might want to ensure that there is not only enough natural shade but that there is some type of additional cooling unit. Many homesteaders will freeze large containers or bottles with water overnight and then put them into the shaded portion of the coop or the run underneath the coop in the hottest part of the day.

Goats and Sheep

Goats and sheep usually prefer to have a pen even if they don't use it all the time. Having a pan with a windbreak and a roof can make the winter and summer seasons much easier on these animals. You'll also want to ensure that they always have access to drinkable water, and if you live in a climate where water will freeze through in pipes, you may need a small heater for their water container. It's also important to ensure that they always have access to food, but for emergency preparedness, they need double what they typically eat throughout the wintertime.

With both goats and sheep, you'll want to have fencing around a very specific pasture, and that pasture should not have all of your livestock together. Goats and sheep can live together fairly peacefully. However, neither of these should be mixed with your cow pasture. Ensure that your pasture has some sunny spots and a shaded area.

There is a special note with goats that doesn't apply to sheep. Goats are destructive. Given any opportunity, goats will go into an area of the land or even your home where they should not be and destroy virtually anything they can. What you need for your fencing is a woven wire or "no-climb" horse fencing that is at least four feet high. Remember that goats are excellent climbers and can jump fairly high, so keep any structures within the pasture well away from the fence line. Putting in and upkeeping this fence is part maintenance and part prevention in order to avoid having to do more maintenance than necessary.

The good news about these destructive qualities is that they turn the ground better than most professional equipment can manage. You should plan, as part of your maintenance calendar, to rotate your goat pasture every after every-other crop rotation.

Cows

It might initially seem that cows are low maintenance animals, but that's not quite the case. Cows, in addition to their food and water, need salt and minerals in the form of licks, they also need to have a spacious pasture. It's also important as part of your maintenance that you spend time with the cows to reduce possible aggressive behavior. Cows are not inherently aggressive, but they are very large animals, and if they get scared or feel threatened, they can do a lot of damage to your property and cause severe injuries as well.

Cattle certainly need shelter, but it doesn't have to be over the top. If you have a wind wall and a roof, or a 3-sided structure, then that is enough for them to seek shelter during poor weather.

Cows do need trimming, and you should schedule a visit with your veterinarian every six months. The veterinarian can come to your property and conduct any trimming necessary, and they can also address any infection that may have taken root in their hooves. Additionally, cows need to be vaccinated regularly for rabies and a wide variety of other contagious diseases. Some of the vaccines will vary by region, so it's important that you discuss these vaccines and a vaccine schedule with your veterinarian.

How to Make Your Homestead as Low Maintenance as Possible

The best thing to do with maintenance is to make it as systematic as possible. That means relying heavily on automated systems and a calendar. For example, with your livestock, you may have a standing appointment with your veterinarian every six months. Having that recurring appointment takes away the stress of having to call and schedule an appointment every time you think that your cow or goat might have a hoof problem.

Low maintenance gardens are often seasonal. A seasonal garden allows you to put fresh mulch and liven up the soil between seasons; however, it does deplete the lifespan of your plants. You can also make your gardens low maintenance by implementing a watering system that will save you from watering them by hand.

Finally, use any resources that you can to ensure that cleaning up the livestock environment is as easy as possible for chickens that can be using linoleum flooring instead of wood. For cows, it can mean having their compost pile right near their pasture, so it is easy to add that manure to the pile. You might also invest in equipment that makes it easier to pick up and transport manure.

Make a Homestead Maintenance Calendar

Over the years, and as you develop your homestead, your homestead maintenance calendar will take shape. For now, you can start with a basic calendar that lays out what you should be doing from month to month. Then after your first-year homestead year, you can look back at the calendar and see what did or did not work out well for you. Every homestead is unique, and your environment and weather conditions will drastically change what your maintenance calendar will look like. However, a starter calendar might look like this:

- **January**
  o Make goals
  o Calculate expenses from last year
  o Sow indoor seeds
  o Clean and disinfect chicken feeders and waterers
  o Ensure all heat lamps are working
- **February**
  o Design spring vegetable garden
  o Tap trees
  o Prune orchards
  o Sow indoor seeds for spring garden
  o Harvest end of winter garden
  o Clean out any pasture area

- **March**

  o Sow indoor seeds for late spring – peppers, tomatoes

  o Change chicken bedding and clean nesting boxes

  o Purchase feeder pigs or make livestock purchase of goats/sheep

  o Maintenance check on roofing and structures

- **April**

  o Plant perennial beds – prepare with heavy compost week before planting

  o Early vegetables

  o Start spring compost heap

  o Change chicken bedding and clean. Nesting boxes

  o Clean and disinfect chicken feeders and waterers

  o Install screens

- **May**

  o Plant annuals

  o Plant late-spring vegetables

  o Clean out any pasture area

  o Introduce new chickens

- **June**

  o Start haying for the winter supply

  o Change chicken bedding and clean nesting boxes

  o Harvest and preserve berries

  o Harvest early-spring produce – freeze vegetables

  o Sow for early fall crops

- **July**

  o Clean and disinfect chicken feeders and waterers

  o Canning

  o Freeze or pickle half of harvest

  o Late haying

  o The seed for late fall crops

  o Mulch garden

- **August**

  o Change chicken bedding and clean nesting boxes

o Butcher broiler chickens

o Freeze or can orchard harvest

o Harvest corn, wheat crops

- **September**

o Maintenance check on roofing and structures

o Clean and disinfect chicken feeders and waterers

o Preserve tomatoes

o Mulch garden

- **October**

o Gather firewood

o Can applesauce

o Can pumpkin butter

o Winterize perennial garden

o Heavy mulch all gardens

o Change chicken bedding and clean nesting boxes

o Put up and check all heat lamps

- **November**

o Freezer baking

o Butcher second round broiler chickens

o Mulch winter beds

o Manage winter garden

- **December**

o Update maintenance calendar for the upcoming year

o Estimated taxes

o Butcher cow, pig, sheep, or goat

o Check pipes and waterers for freezing

Maintenance can be rather easy if you plan well but there is really no telling what lies in store for your first year. Using a generic calendar like this one where you simply remove any tasks that don't apply to you is a great place to start. However, you may need to make adjustments as you go. It can also be difficult to start maintenance as you're still developing your homestead and making your homestead plan. Unfortunately, maintenance is hard to catch-up on. So, if your garden is up first, then start your garden maintenance the very next month. Then start maintenance on your chicken coop the month after your chickens get moved in. Then start maintenance on your harvest when your first harvest comes around. That can help you build a maintenance calendar while you're still setting up your homestead, and it can make maintenance down the road much easier.

Chapter 11: Sharing Your Learning Experience

One of the unique things that come with homesteading is the ability to share your experience. Although you are far from being the only homesteader around, the challenges that you face may not be the same challenges that others have experienced already. By sharing your experiences, you can help others better prepare for changes in seasons, issues with crops and animals, and much more. However, it's not just about giving back to the homesteading community. You can even create another avenue of income by sharing your experiences.

Through blogs and community events, you can become a teacher in your own right. We'll explore each way that you can share your experiences, and the benefits and possible revenue that could come with each option. If you decide to share your homestead experience through a blog or events within your community, you'll need to consider the skill set you already have and your ability to market yourself and your homesteading brand.

Vlogs

Vlogging or video logging is the method of recording your experiences and posting them online in a video format. Typically, the people that are doing this are just called YouTubers because YouTube is the most popular platform at the moment. However, people were creating video diaries and recording certain aspects of their life long before YouTube existed.

Don't get stuck asking yourself if you can be a YouTuber. Instead, ask yourself if you can approach documenting your life at an almost nonstop pace.

To start a vlog, you should consider a few of the main elements. For example, because of YouTube's presence, you will likely need a YouTube channel. However, to reach other audiences, you might also consider having a website. In addition to a website, you may reach out to followers, fans, or people searching for content through social media platforms such as Instagram. One of the major mistakes that people new to vlogging make is that they pigeonhole themselves into one platform. Then when another platform takes off and gains attention, they can't pivot in time to keep up with all the changes.

Essential Elements of a Vlog

- High-quality documentary-style videos
- The ability to present yourself in front of the camera well
- Basic video editing skills
- Recording equipment
- Time to dedicate to posting and maintaining your channel

Now, you don't need to start out with all of these things. In fact, many people start by recording videos on their phones or learning to video edit as they go. There's no need to rush out and spend thousands on video recording equipment if you're not even sure if it will work for you. Basically, if you're not passionate about it, don't worry about it. Vlogging can turn some people away from their passions, such as homesteading. It's common for "YouTubers" to experience burn-out after a few years and then completely abandon the things that they wanted to share on their channel.

When it comes to revenue, there is a lot of opportunity with vlogging, if you're willing to put in the work. You probably won't become an overnight celebrity sensation that regularly receives tens of thousands of dollars in advertising revenue from YouTube. However, you might be able to create a nice supplementary income. It's important to note that vlogging does not generally turn out as a consistent income. While top earners on YouTube make upwards of $10 million annually, you shouldn't expect to hit those numbers. In fact, when Learning Hub sat down and averaged out the views per monetized video, you'd need about 1,000 views on any particular video to earn less than ten dollars. Technically, the advertisers working through YouTube's monetization system payout $0.18 per view or $18 for 1,000 views but Google, the ad aggregate for the platform keeps 45% of that revenue as a type of finder's fee —leaving the vloggers with only $9.90 per 1,000 views. There are other elements to consider, and advertising through YouTube's monetization system is not the only way to make money as a vlogger. Some of YouTube's top influencers or personalities don't use YouTube's monetization system because they don't want to interrupt their viewers with ads. Instead, they may align their personal brand with a sponsor or work with other local brands to advertise in a non-invasive way. A lot of people who craft, homestead, or do other lifestyle-esque videos will choose to do this as it allows them to give information and valuable input on brands and products within the industry.

So, what happens if you don't want ads, and you don't want sponsors? Well, you can still use vlogging as a platform to share your experiences without it producing an income. Or, you can create your own merchandise that promotes your brand. You can also add in a system such as Patreon, where you can ask subscribers to pay a membership fee to access additional or premium content in addition to what you already post. Often vloggers that do this will create fun and entertaining videos for free, then use Patreon or a similar platform to provide tutorials, how-to videos, and more. When doing this, you'll need a bit more tech for proper execution. You may need a mailing list and website to make sure that your premium subscribers get all the additional content that you promised.

Is it worth it? For many people, the act of vlogging is relaxing and cathartic. It's a creative outlet to share this monumental change in your life. Most people haven't grown up homesteading, and it's probably safe to say that even the experienced can have unexpected situations, and such things crop up as they build their homestead. If you start your vlog channel as you build your homestead, or even during the planning process, you're cataloging or recording your experience not just for others but for yourself as well. This is a unique way to share your experiences and get involved in a community that you may not have known even existed.

Blogs

Blogging is exactly like vlogging, but instead of using a video format, you deliver written content. If you love to write or journal, then blogging is an excellent way to document and share your homesteading experience. Typically, through blogging, you'll post to one platform, and it will read as a blend of an article and journal-style entry. It really won't read like a news story or an article that you might read on social media platforms; it should be more conversational in structure. Many people write their blogs as though they're talking to a close friend. That allows the language to be a bit more flexible and stray from rigid structures that you might expect to see in most textbooks or guides.

It is important to note that you don't need to be an all-star writer to have a blog. In fact, while having a general grasp of grammar is always good, a lot of blogging breaks standard rules for writing. Because it's a conversational tone in approach, blogging is a lot more forgiving with writing fundamentals than starting a novel or writing a weekly news article. If your first instinct is, "I can't write, so I can't write a blog," then give yourself a little credit and try anyway.

A few of the blogs that stand out among homesteaders are The Self-Sufficient Homeacre and The Homestead Survival. They use two very different approaches to blogging in tone and style. They also have different scopes on what they share regarding homesteading.

Looking at The Self-Sufficient Homeacre, a lot of the articles have to do with growing plants, raising chickens, preserving the harvest, and cooking. It's a pretty typical look at homesteading. One article entitled, "How to Grow Food in Small Spaces" delivers an onslaught of options for growing plants when you're tight on space. The tone is fairly direct but still friendly, and the website itself is fairly "clean" but does contain a few posted ads and then affiliate ads.

The Homestead Survival has too many ads, and the webpage or blog looks pretty cluttered. However, everything is organized and has a place, and is generally easy to navigate. What they have done here is create many smaller options to access sections of their website which address different needs. You're not scrolling through endless articles that flit from one topic to another. This blog delivers household tips, DIY projects, home remedies, food storage and canning tips, and even knife skills for those who raise animals for meat. The Homestead Survival is a drastic contrast to the other blog, and the tone is a lot more personal. They are transparent about their numbers in terms of what they spend and what they earn, and whether a DIY is cheaper than buying something pre-made.

The point in comparing these two blogs is to show that the content itself can vary drastically. Both listed here are homesteading blogs; both generate revenue for their website owners. However, it's important to know exactly what you want to deliver through your homesteading blog. A lot of time and dedication goes into creating and then maintaining a blog, so if you're not confident in handling daily tasks that may or may not generate revenue, blogging may not be right for you.

To Start a Blog, You Will Need:

- Website or platform such as Tumblr
- Plenty of content – It's best to start a blog with at least 50 articles ready to publish.
- Time to continue writing articles to support your blog; consistent posting is key!
- The domain name and web host (which can quickly become expensive)
- Strong social media presence

Unlike using a video platform such as YouTube, you don't have access to a giant base of people. You'll need to pull traffic to your site. To generate traffic to your site with the hope of someone clicking an ad or purchasing an affiliate product, you will need to have a really strong social media presence. You may very well end up blogging and vlogging to generate the necessary income that you expected from sharing your experiences.

Many people choose to blog, however, without ads or without demanding an income from this creative outlet. If you're looking for a way to reach the homesteading community, a blog is a good option. Not everything that you do at home needs to generate an income. However, there are ways to provide an ad-free user experience or to use a free platform that you can't monetize and still profit from your blog. Going back to Patreon, you can market premier content, merchandise, and even one-on-one lessons while still hosting a free blog that serves as a creative outlet for you.

Hosting Community Events and Getting Involved

Hosting a community event is a huge deal. Fortunately, a lot of the soft skills that you've developed through setting up a homestead will pay off here. All that work you put into planning the homestead will feel pretty familiar when you're planning your community event. You may request permission from your city to host an event for your neighborhood or host a private event with a limited capacity.

Each option has different requirements when it comes to permits and planning, so it's worth the effort to decide before you even start just how big your event should be. It's always best to start small and work your way up. There is no need to have a citywide event if you don't know the extent of interest within your community.

Private Event with Set Capacity

Now, in theory, you could host a private event at your house and not have to request any permission from your local government. It would be about the same as having a barbeque over the weekend. However, if you're going to be selling goods, providing food, or taking up a substantial amount of street parking, it's better to alert the city of your plans than to face a possible ticket later.

As you plan a private event, you'll need to put a cap on how many people can attend. You could always start with a shortlist of people you would like to invite and ask them if there is anyone that they would like to bring, or leave it open on your invitation list with a plus-one for each person. Hosting private events like this are a good way to get a feel for how many people are interested in homesteading. People like your siblings or close friends might show up just to support you, but it can still be a fun and informative experience for them and whoever they bring with them.

When you do a private event, you might want to focus on one primary element of homesteading. To cover all of homesteading in just an hour or two would be overwhelming for you and your attendees. However, if you do a private event on herb gardening, you can better control the conversation and keep people engaged rather than jumping from herb gardening to canning to raising chickens.

Keep in mind that when you host a private event, it's always best to do everything that a host should do. That means make sure that parking is available, having food or some type of snacks available as well as beverages. Make your guests comfortable, and it can be a great experience for everyone involved.

Neighborhood Events

Neighborhood events can be really fun, and they can become a monthly or an annual event. If your neighborhood doesn't currently do block parties, then you might not know where to start. First, canvas your neighborhood and find out how many people are interested. By this point, it might already be well known that your lemon tree is the best place to get lemons in town. So why would your neighbors miss the chance to talk to you about home-growing plants and the differences between what you produce at home and what you find at the grocery store?

When you're implementing a neighborhood event, you should expect a lot more people than you might have at a private event. This gives you a little more opportunity to address different elements of homesteading and cover a broader range of topics than just the one thing people wanted to talk about. When planning a neighborhood event, you should collaborate with other neighbors as well as the city to ensure that you have permits for everything and that there are enough activities and information centers for everyone to be entertained.

One of the more important elements that you need to express when you're doing a neighborhood event is that you're available to help people that want to get started. Sharing your experiences is the first step in helping other people understand and maybe begin homesteading. Ultimately, when you're sharing your experience in a face-to-face situation, you should be open to getting much more involved.

Volunteering Through Community Outlets

Community outlets such as 4-H clubs or the Future Farmers of America are a great way to get involved. These groups and other similar groups connect adults with a certain skill set with teens and children who aim to develop those skills. As a homesteader, your ability to share your experiences can help others cultivate a skill set for building self-sustainability within their lifestyle. Being an active participant in these communities allows for the sharing of your experiences, but it also allows you to engage on a deeper level.

You're not just teaching children how chicks are hatched. You're teaching them about the natural life cycle. You're teaching them about the importance of what chickens do for our world. For many, it's an uplifting way to share your experiences because you're going beyond what just happens on your mini-farm, and you're expanding that to show how people can create the experience that you've had and build something for themselves, too.

Getting involved with community outlets is fairly easy. You can reach out to local chapters of well-established clubs that we mentioned above, or you can contact the city and your school district. Even if your city doesn't have a local 4-H club, it's likely that some of the high schools in your area do have agricultural classes. Because of how the school system has changed, many people in those classes don't have access to an actual garden. It's all theory learning. But by volunteering your experiences and possibly your homestead for a short time, you can connect with teens or young adults who are looking for ways to develop their skill set.

Sharing Through Storytelling

Now that you have an idea of the different methods for sharing your experience, you might wonder what comes next. What stories do you have to tell? Is homesteading too boring to share with others? Is it too much like what everyone else is already doing?

One of the reasons why we touched on two of the most popular blogs in the homesteading world is to show their differences. Open up any two homesteading vlogs on YouTube and you'll get drastically different accounts of homesteading, because everyone's experiences are unique. You may have suffered from the same cold snap as a fellow homesteader in your area, but there's no way that you and that person experienced it in the same way, responded in the same way, or came up against the same hardships with your plants and animals.

One of the best ways to start sharing your experiences, no matter your platform, is to use storytelling. Storytelling is something we do naturally and is a great way too not only teach but identify with people that you're sharing an experience with. For example, you can set out a list and say in the event of a hard freeze, you can do X, Y, and Z to sustain your crops. However, that doesn't communicate your experience. Through storytelling, you can say something like this:

"Winter came in, but I wasn't too worried, because it was the desert. My homesteading experiences would be harshest in the summer when I had to worry about keeping everything alive in the heat. But, when winter came, I realized that the desert gets exceptionally cold. I wasn't prepared with firewood or extra batteries, and I hadn't planned on things like heating lamps at all. After the first night, my chickens were in a sour state after having gone all night without water, and my poor herb garden was frozen. I went out the next day, get a small water heater for my chicken's water, and laid down some electric heat cable through the part of my garden that survived that first brutal night."

Simply put, posting a list or a how-to doesn't really help the community or help you explore your experiences from a reflective standpoint the way that storytelling does.

When sharing your experiences, keep in mind that people are listening, they're interested, and that you have something important to share. Your experiences can help others build better homesteads or even encourage them to try living a more self-sustaining life on their own. It's something that every homesteader should consider doing because of how much sharing your experience can contribute to the homesteading community. It can also offer quite a bit of relief on your part. You can look back and see what seemed like a monumental struggle at the time, and now in retrospect, it seems simple. You realize it was something important that you needed to learn and drastically changed your system for homesteading. Sharing most of your experiences can help others better prepare and give insight into your unique solutions to various homesteading challenges.

Chapter 12: Expanding Your Homestead

There are very few homesteaders that remain happy doing just the basic maintenance and upkeep on their land. Most have that drive to continue developing and cultivating new skills. Don't be surprised if, after a first year or two, when everything seems easy, you find yourself wanting to add something into your homesteading mix.

Fortunately, there is a vast array of homesteading specialties that you can add to your land. Some of these specialties don't necessarily fit the modern backyard homestead model, but if you have extra space, there's no reason you can't explore all of these more advanced homesteading activities.

Expanding into Cattle, Goats, Sheep, Pigs and Working Dogs

Raising livestock is difficult and you need to carefully consider how and when you will introduce them. Typically, goats and sheep need little room at all. Goats on their own don't need much space, but they can get destructive when they don't have enough room to move around.

You want to be sure that you have any necessary structure and pasture ready before you purchase any more livestock. When you expand into livestock, you have a few options about how they will contribute to your homestead. Cattle, goats, sheep, pigs, and even rabbits can all contribute to the homestead's meat supply. Additionally, cattle and goats are excellent for milk production. If you want to raise livestock, you can also explore options for herd-sharing and stud services to have your livestock bring in a profit.

If you're going to be adding livestock to your homestead, you should seriously consider bringing in a working dog as well. While it is always nice to have a dog around the house, a working dog's top priority won't be friendship. Well raised working dogs will have a constant need to be close to their herd, and they serve as a form of protection and stress relief. Cattle, goats, and sheep can all benefit from a working dog. A working dog can help the herd focus. It can keep them company, it can make them feel a sense of security, and it can alert you if anything is wrong at any time of day or night.

Some of the best breeds of working dogs include the Bernese mountain dog, Great Dane, German shepherds, Australian shepherds, and bloodhounds. These breeds are all athletic and should, with proper training, get along with nearly all other animals. Additionally, most of these breeds don't have a problem attacking larger pests such as raccoons or opossums that may bother your herd.

Beekeeping

As a homesteader, adding bees to your property can ensure that you always have a natural sweetener at home. And the bees can help pollinate your gardens, meaning larger yields. There really is no downside to beekeeping except maybe the occasional sting. To get started, you should reach out to your local beekeeping organizations and learn how bees use their hive. You'll want to have their hive and all of your necessary materials on hand before you get the bees on your property. It is always best to start your hive in the spring so that your bees can immediately get to work.

But you'll also have to learn how to harvest the honey, keep their hive clean, and store honey properly. Beekeeping can open a lot of doors in terms of creating products using beeswax and using home-gathered honey in your cooking. Bees don't require a lot of space, and you don't need a large hive, so you can start beekeeping with only a small amount of space.

Aquaponics

Aquaponics is a combination of gardening fish and vegetables together. Fish produce waste that feeds microbes, and then those microbes feed the plants with nutrients, and then the water gets filtered by the plants and returns to the fish. It is a complex system and does require quite a bit of careful planning. Additionally, you'll have to maintain the water constantly to ensure that your fish are healthy, and your plants are safe. But the payoff is that you can have more nutrient-rich vegetables and the availability of fish right in your backyard.

To get started with aquaponics, you might consider just having a fishpond. Then you can begin the introduction of aquaponics with controlled steps. The fish in a fishpond do need some care, but generally, they are self-sustaining. You may need a pump to keep the pond clean, and you may need to feed the fish regularly.

Hydroponics

What if that garage or shed you don't use could convert into more garden space? Hydroponics is the way to grow plants without soil. Typically, with hydroponics, plants will grow faster and offer bigger yields. Additionally, you can save a lot of space because instead of relying on nutrient-dense soil, you're giving the plants oxygenated nutrient-rich water.

Although hydroponics requires a lot of research to understand and implement, the basic approach is to keep plants' roots wet with nutrient-dense water solutions. You'll also want to ensure that you're providing the proper type of lighting for the plants that you're growing. Most plants that you would have in a garden require direct or partial indirect sunlight. So, in addition to having the hydroponic system set up, if your hydroponics garden is indoors you'll also need a full-scale lighting system.

Vertical Gardens

Have you run out of space but not great ideas for vegetables to plant? It is possible to grow vertical gardens where the plants are set into a pot that hangs off the side of the wall. You can mount vertical gardens on the side of your home, shed, or even your chicken coop. Some people worry about the extent of water damage that could happen to the side of their home or shed, but you can always protect against that with a bit of planning.

These vertical gardens are grabbing a ton of attention because of how simple they are to make. Just be sure that you have the time to care for them.

Expanding Your Homestead

It's amazing how much starting a homestead can boost your confidence in your ability to develop new skills and try new things. You may try a wide variety of expansions to your homestead, such as growing an herb garden or exploring different methods of plant cultivation. Expanding your homestead doesn't always call for more space. Always remember to be creative and use the resources that you have available. Homesteaders are often experts at finding new creative solutions to achieve their goals. Celebrate your homesteading anniversary by setting new goals and planning to build new skills!

Conclusion

Homesteading is a rewarding experience. You have learned how to plan throughout the year, accommodating different seasons, and living off the land. You will also face challenges that may demand you improve your skills and abilities. Homesteading is about self-sustainability while still devoting yourself to cultivating the land you have available to you. It is the opportunity to learn how to live simply while developing skills and your personality in a way that probably never seemed possible in our technological age.

Use the methods for planning, preparing, growing, and nurturing in this book to help map out your first few years as a homesteader. Then push yourself further. Share your experiences and help others develop skills that can help them become more self-sustaining. Homesteading is a lifestyle that brings joy and relief into your life while filling your days with tasks and hobbies that pay off day after day with outstanding physical and emotional rewards.

Part 2: Backyard Chickens

A Comprehensive Guide to Raising Chickens for Beginners, Including Tips on Choosing a Breed and Building the Coop

Introduction

Self-sufficiency, especially when it comes to your food sources, can be extremely liberating. No one wants to spend hours reading labels trying to figure out whether or not something is healthy. While you may not be able to control where all your food comes from, raising backyard chickens is one way to make sure that you are getting healthy and wholesome eggs from your own flock.

Chickens are pretty low maintenance, and this is probably the reason raising backyard chickens has become a popular hobby. After all, who would not want a pet that also provides fresh eggs? However, it is not all about the eggs. Chickens are not just a source of eggs, but they also make interesting family pets that your whole family can enjoy.

While you may have an interest in raising chickens in your backyard, you probably do not know where to start or even how to do it. That is why we have compiled a simple and straightforward guide for beginners who want to raise chickens. Some people may have no experience at all with raising chickens, but that should not deter you from taking up this rewarding hobby.

This book gives you detailed information on how to start your backyard flock. We want you to be able to take care of the chickens once you have them, so this book also gives you in-depth insights into how to raise, care for, and maintain your backyard chickens. From building a coop to equipping it will all the accessories you need to take proper care of your chickens, you will get all the information you need to get started.

A significant part of the book is dedicated to the proper care of chicks and how to raise them from day-old baby chicks to adult hens that lay eggs. We understand that caring for pets is a delicate process, and this book aims to equip you with all the knowledge you need to raise a healthy and happy flock.

We have a section devoted to understanding chicken behavior that is geared to help you bond better with your flock. That part of the book is designed to help you pick up on any distress signals in your flock or any signs of illness. If you have thought about raising backyard chickens for a while, this is the book that will guide you on how to do it. You need not be a farmer in a rural setup to raise healthy chickens. With the right tools and information, you can transform your backyard into a safe haven for chickens.

Naturally, the first place to start on your journey to raising backyard chickens is understanding why you need to do it at all.

Chapter 1: Why Raise Chickens at Home?

Raising chickens is no longer just the preserve of people with rural farms. An increasing number of people are turning to raising chickens in their back yards. You have probably thought about it yourself, but, like any other venture, you may still wonder whether the pros outweigh the cons. If the allure of fresh eggs is not enough to convince you, there are still plenty of reasons to raise chickens in your backyard.

Whether you are looking for a fulfilling hobby or, like many others, just want to have more control over what's on your dining table, chickens are a great place to start. For most people who consider raising pets or livestock of any kind, space is usually one of the main deterrents. The beauty in choosing chickens is that they do not take up a lot of room. If you have a small backyard, you can still comfortably accommodate a small flock of chickens. For a modest flock of about six chickens, you would need approximately 110 square feet of space. This is why more and more people have taken up raising chickens in their backyards. With just a modest amount of space, you can have your own supply of fresh eggs readily available and a rewarding hobby to boot.

Another key concern that you may have when deciding to raise chickens is how much time and maintenance is required. Pets come with their fair share of maintenance, and chickens are no different. However, chickens are pretty low maintenance, and this is not one of those pursuits that will take up hours of your time. Keeping chickens will require some effort on your part, but for the most part, these birds are pretty self-sufficient and do not require round-the-clock care.

As long as you have a secure coop to keep your flock safe from predators, you will find that chickens require little attention. Most people find that the daily care required for chickens takes up less than half an hour a day, so this is not one of those hobbies that will turn out to be time-consuming.

If you have a dog or a cat, you will realize that chickens are easier to care for than either dogs or cats. Chickens do not need as much human attention as other household pets. Provided they are well fed and housed, you will find that you can pretty much go about your business without needing to keep checking on your chickens all the time. Most people who keep chickens will tell you that they are easy to keep because they are pretty independent.

For people with children at home, getting pets that your children can be around safely and also help care for is always a plus. As far as pets go, chickens are mostly non-territorial, and most are not aggressive. This means they make great pets that your children can also enjoy taking care of. Nothing teaches young children about responsibility better than having them help out in caring for a pet.

Chicken-watching can also make for a fun and interesting pastime. Chickens have individual personalities and quirks that make them fun to watch. They can also be beautiful, and depending on the breed, some can be unique in terms of appearance. Some chickens are also friendly and will approach you or your kids whenever you are within their sight. This means this rewarding hobby will be anything but dull.

If your backyard is well enclosed, you can let the chickens roam free during the daytime since chickens tend to be pretty friendly and will not be aggressive. Of course, if you are to let your chickens roam free in the backyard, you need to ensure that they are safe from predators.

While having a pet is great, chickens come with the additional benefit of being a source of fresh food in terms of eggs and meat. While you can buy eggs at the grocery store, you can never be sure about the freshness or quality of store-bought eggs. The increasing popularity of raising backyard chickens, even amongst celebrities such as Jennifer Aniston, Martha Stewart, and many more, is partly because people are getting more conscious about the kind of food they eat. When you get your eggs from your own chickens, you know what they have been eating, and you are, therefore, in control of what you are eating. This means you can choose to feed your chickens organic food and, as a result, enjoy organic eggs that are nutritious and free of GMO additives.

When you buy products from a grocery store, you have little knowledge of what kind of produce you are getting or how the chickens that produced the eggs were raised. This means that you cannot be 100% sure about the quality or freshness of the eggs you are getting. But with your own flock, you have control over what they eat, and you gather up the eggs daily, so freshness and quality are guaranteed. This also goes for people who want to keep chickens for meat. With your own chickens, you are assured of quality and that what ends up on your dining table is free of any harmful chemicals and additives.

Research studies show that eggs from free-range chickens tend to have higher concentrations of nutrients such as beta-carotene, omega three, and vitamins A and E than eggs sourced from battery hens or chickens that are raised in cages. So, if you have been on the fence about raising your own chickens, consider the additional health benefits for you and your family that come with having control over the kind of eggs you eat.

Eggs and chicken meat can also provide you with an additional income stream. A lot of people prefer organic produce, and you can easily find a market for any extra eggs that your chickens produce. This means that other than feeding you and your family, chickens can also serve as a source of income, and ultimately you may find that the chickens end up paying for their feeds and maintenance cost.

If healthy nutrition sounds good to you, but you are not sure about the kind of costs that will come with raising backyard chickens, that's another area where chickens have an advantage over other types of pests. Chickens are relatively inexpensive to acquire and maintain. For most breeds of chicken, the buying costs per bird will generally range from as low as $3 to $30. This cost will depend on factors such as age and breed, but all things considered, for a pet that is going to provide you with eggs and meat, the startup cost for raising backyard chickens is pretty minimal.

As for chicken feed, this is another area that will not cost you too much. People with modest flocks of six birds or less find that chicken feed only sets them back between $20 and $30 a month. Chicken feed tends to be inexpensive and readily available in feedstores, pet stores, and even grocery stores. There will be differences in the cost of feed based on the brand you choose and the type of feed you go for, but on average, most people find that chicken feed is affordable.

Another plus when it comes to feeding chickens is that they are omnivores and therefore tend to eat most types of foods. Besides the chicken feed, you will find that chickens will happily eat any scraps off your table. This means that any scraps or leftovers from your dinner table need not go to waste.

Chickens are also great foragers. When left to roam in open areas such as yards or gardens, they will dig up bugs and other types of edibles that they find on the ground. Chickens are not fussy eaters, and this means that you do not have to worry about keeping them well-fed and happy. Of course, it is important to ensure they eat healthy stuff since you want them to produce high-quality eggs.

For people with gardens, chickens are not just inexpensive, low maintenance pets, but they also make some of the best natural manure. If you want to fertilize your garden using organic manure, chickens will provide you with one of the best natural fertilizers. Poultry manure contains plenty of nitrogen, phosphorous, and potassium, as well as other essential nutrients that improve the quality of your soil and give you healthier plants.

Since organic manure is safer for the environment, your crops, and your health, you can simply clean out the chicken manure from the coop and add it to your compost heap. Additionally, if your chickens are free-range, they will effectively fertilize the soil in your garden or yard as they roam about. Since chickens also tend up to dig and scratch the ground for bugs and eat unwanted weeds, they make great garden tillers, especially when you are getting ready to plant or have just cleared your garden.

When you want to get your garden ready for the next crop, your chickens can help with the clean-up and fertilizing of your garden. By using the organic manure from chickens instead of artificial fertilizers and other chemicals, you will find it much easier to grow organic crops in your garden. The kind of products or chemicals that you use in your garden will end up in your plants, and ultimately in your food, so organic manure gives you the option to grow food that is organic and free of harmful chemicals. If you are looking to farm more organically and reduce the use of artificial chemicals and fertilizers in your garden, you will find that chicken manure is an excellent inexpensive alternative.

Chickens may just be one of the most useful pets that you can keep in your backyard. These handy birds are good for more than just eggs and will prove a worthwhile addition to any backyard. If you have thought about raising your own chickens for a while, with a little bit of effort on your part, you can enjoy a host of benefits from your flock. While you will need to put in some effort to raise chickens in your backyard, you will find that the rewards will far outweigh any cons.

Chapter 2: Things to Consider Before Getting Chickens

There are plenty of reasons that make raising your own chickens a pleasurable and rewarding venture. From having a ready supply of fresh eggs to the pleasure of watching your chickens thrive, raising backyard chickens is appealing on many different levels. However, as much as keeping chickens is rewarding, it is still a responsibility that requires time and effort.

This means that before you join the bandwagon of people raising chickens in the backyard, you must be sure you are up to the task. There are important things that you need to keep in mind before you embark on raising chickens. Let's take a look at some of the factors you need to consider.

I. Are chickens allowed where you live?

II. Do you have enough space in your backyard to raise chickens?

III. Do you have the time to raise chickens?

IV. For what purpose are you raising chickens?

V. Are you ready for the costs involved?

VI. Do you have other pets, and if so, can they co-exist with chickens?

Are chickens allowed where you live?

If you are a beginner to raising chickens, before you consider any other factors you first need to ensure that you are allowed to keep chickens in your area. This means you need to find out what the local laws are. The last thing you want is to get on the wrong side of local ordinances and laws.

To find out whether you can legally keep chickens where you live, you can check with your local zoning office or county office. Most towns and cities will have their own regulations on keeping livestock and poultry. Some counties also provide online resources that offer guidance to people looking to keep poultry or other livestock in their backyards.

You may find that you are required to acquire a permit for your chickens. This is more or less similar to the kind of permit you get for dogs or cats.

You may also find that, even if the laws in your area permit the keeping of chicken, there is a limit to the number of chickens you are allowed to keep. This limit will depend on factors such as the size of your land and property lines. However, each county has its own set limit on the number of chickens you can keep. Once you have this information, you will be able to comply fully with the set regulations and avoid any complications down the road.

In some places, the local ordinances are flexible, and you may be able to get a permit for keeping extra birds above the stipulated limit. Another regulation that you need to familiarize yourself with is whether or not you are allowed to keep roosters. Roosters tend to pose noise concerns, and the keeping of roosters is not allowed in some towns and cities. Some areas will allow you to keep a rooster but only up to the age of four months.

You will also need to understand whether your local ordinances permit you to have chickens that roam free in your backyard. Depending on where you live, you may find that there are enclosure restrictions that require you to keep your chickens in an enclosure or within a contained environment. These will be an important regulation to seek clarity on, especially if your aim is keeping chickens as a free-range flock.

In some areas, you may need to get approval on your coop plans and building materials before you can set up a coop in your backyard. Before you start building your coop, check your local laws to see what the regulations are. In some cases, you will find that there are distance regulations imposed to guide you on how far your chicken coop should be from property lines. The distance required from property lines can range from 10 to 90 feet, so make sure you are clear on what your local law stipulates before putting up your coop.

In addition to local laws and ordinances, you will also need to check if there are any regulations on keeping poultry made by your neighborhood residents' association. Since chickens can be a noisy, smelly, hygiene concern for your neighbors, it is always advisable to check for any neighborhood laws that regulate if or how you can keep chickens. You do not want to aggravate your neighbors, so giving them a heads up on your project may help to ensure some goodwill and prevent resistance from the people who live in your area.

Ultimately the laws in your area will determine whether or not you can keep chickens, how many you can keep, and any other regulations. If your local town or city does not allow the keeping of chickens, that does not necessarily spell doom for your dream. People have successfully petitioned their local governments to change their ordinances and laws on the keeping of poultry. You can do this through your local city council. Changing local ordinances or laws may take some time, but if you are patient and consistent, you may end up having the regulations in your area reexamined.

Do you have enough space in your backyard to raise chickens?

While chickens take up relatively little space, you still need to make sure that your backyard has enough room to accommodate a chicken coop and run for your chickens. The general rule of thumb is that you need at least three square feet of space per chicken in your chicken coop. This means that the larger the flock you want to keep, the more space will be required.

Your chicken coop needs to have enough space for the chicken feeders, water containers, and a nesting box as well as a roosting area where the chickens can perch. Chickens spend a lot of time in their coops, so it is important to ensure that it provides a safe and comfortable space for them. When the coop is too small, the smaller chickens may get bullied by the bigger ones. Also, bear in mind that you need to be able to get into the coop to clean it and gather your eggs, so you need to make sure that there is enough room in your chicken coop to stand and work in.

Chickens will also need a run. This is the space on which they can roam and forage. On average, a run of at least fifteen square feet per chicken is adequate, though if you have more space that would be even better. When chickens have ample coop and run space, they are less likely to infect each other with diseases and parasites. Just like you wouldn't want to have any other pet cooped up in a tiny space, ensuring your chickens have enough room is crucial.

If you are looking to raise their chickens as free-range without a run, keeping them in containment, the bigger the space that you have for your chickens, the better. This means that on average, you should work with about 25 square feet per chicken. However, always bear in mind that if your chickens are allowed to roam free in the yard, you need to have safeguards protecting them from predators.

On the whole, chickens will not be too demanding in terms of space. However, before you embark on raising your chickens, set aside the area on which you want to raise your chickens. The size of this area will then guide you as to the number of chickens you can comfortably accommodate in your backyard. Chickens that live in spacious and well- designed areas are ultimately healthier and happier.

Do you have the time to raise chickens?

Having a pet is a responsibility, and chickens are no different. To avoid getting caught up in a hobby for which you are ill-prepared, it is important to understand the kind of responsibilities involved in raising backyard chickens. While keeping chickens has more than its fair share of benefits, there are also chores to contend with, and anyone looking to raise chickens has to be willing to put in the time required.

Most people who keep backyard chickens will tell you that chickens are easy to maintain and do not require round-the-clock attention. However, they do still need to be fed and watered daily, their coops need to be cleaned, and, of course, you will need to collect the eggs. This means that you need to allocate time daily for feeding as well as collecting eggs from your chicken coop.

While thirty minutes a day may not be too taxing for most people, if you travel a lot or are away from your home for extended periods, you will need to have someone taking care of the chickens while you are away. Naturally, the bigger your flock is, the more time you will need to care for your chickens adequately. For beginners, it is always advisable to start with a modest flock. Once you get the hang of the maintenance involved and the ins and outs of caring for chickens, you can then gradually increase the size of your flock.

Chickens tend to poop a lot, and dealing with manure is par for the course for people who raise backyard chickens. This organic manure can get quite smelly if allowed to accumulate, so you will need to find time to clean out your chicken coop regularly. Aim to clean out your coop weekly to prevent the build-up of manure in the coop. Since chicken poop can harbor bacteria such as salmonella, you will need to have protective gear to use when cleaning out the chicken coop. If you have a garden, this organic manure can be used as a fertilizer, so it will also serve a purpose in your garden.

You will also need to clean the waterers and chicken feeders weekly to ensure that your chickens have access to clean, uncontaminated water and food. Thorough cleaning and deep sanitization can be done twice yearly. While several chores will be involved in caring for your backyard chickens, some of these chores only need to be done weekly so they can be easily managed. However, some people do find some of these chores unpleasant, and so before you decided to keep backyard chickens, you need to be sure that you are up to the task.

The health and well-being of your chickens will depend on how well they are cared for. This means that while chickens will provide you with eggs and many other benefits, you will also need to give back in terms of time and effort. Most people get into hobbies without realizing how much time and work will be required, and end up regretting their project. Avoid this pitfall by carefully considering how much time you are willing to spend on raising chickens.

For what purpose are you raising chickens?

People raise chickens for different reasons. Some do it for the eggs, others for the meat, and some do it for pleasure. Whatever it is that you want to raise chickens for will be an important factor when choosing the type and breed of chickens to keep, the size of your flock, and how you choose to raise your chickens.

A chicken, contrary to popular belief, is not just a chicken. There are varied breeds of chickens that each bear unique characteristics. This means that some breeds are better suited for some purposes than others. Chicken breeds vary greatly in terms of temperament, noise levels, egg production capacity, and many other factors.

Some chickens adapt better to confinement than others, so these types of breeds work well for people who are not going to free-range their chickens. Things like noise level and temperament are also important factors to have in mind when determining the best breeds to keep in your backyard.

Chicken breeds that are ideal for people who are mainly keeping chickens for eggs include breeds like Barred Plymouth Rocks and Rhode Island Reds. These two breeds do well as backyard chickens. They are prolific egg-layers and will provide you with a steady stream of eggs. These breeds adapt well to confinement and are generally not noisy, meaning that they will be less of a noise nuisance to you and your neighbors.

Rhode Island Reds also tend to be docile and friendly, so this is a breed that even children can be around and help take care of. In essence, these two breeds check most of the boxes for what you need in a backyard chicken. Another breed that also does well as backyard chickens is the Jersey Giant. It also has a calm temperament. However, Jersey Giants tend to be large and may, therefore, require more space than other egg-laying breeds.

If you want to raise backyard chickens as pets or for pleasure, you will be better off choosing calm and docile breeds such as the Rhode Island Red. Breeds like Araucana are hardier than other backyard chicken breeds but tend to be temperamental and may not make the best pets, especially if you have children. So, before you buy your first flock, always consider the purpose for which you want the chickens. This will help you select the best breed for your needs and avoid disillusionment down the road.

Are you ready for the costs involved?

Raising chickens is a pretty inexpensive venture. However, there are still costs involved, and you need to be ready to inject some cash into your hobby. The initial costs will involve expenses such as paying for permits, buying your chickens, and of course the cost of building a coop and run for your chickens. This means that the higher costs will be at the start of your venture. Once you have everything in place, maintenance costs tend to be largely related to buying feed and getting veterinary care for your chickens if and when the need arises.

At the onset of your project, you will need first to ensure that you have a proper enclosure and housing for your chickens. When it comes to coops, you can buy a ready-made coop or build one yourself. While ready-made coops can save you the time and effort it takes to build one, you will end up spending more money than if you were to build the coop yourself. Online stores such as Amazon have a wide range of chicken coops available that range in price from budget-friendly to costly. This means that you can shop for one that meets your needs but is still within your budget.

When it comes to chicken coops, building your own is usually a popular choice for most beginners. If you are handy with outdoor projects, you can save a pretty penny by choosing to build your own chicken coop. All you need is the building materials and a building plan for your coop, and you are good to go. Some people enjoy putting up their own coops because they can make it exactly how they want it to be in order to best suit their needs.

Another advantage in terms of costs is that when you choose to build a coop yourself, you can easily use recycled material to make the chicken coop. This means that you can make use of whatever appropriate materials you have at hand without necessarily having to buy new building materials. Again, this is a plus if you want a cost-efficient way to start keeping your backyard chickens. Some people find it easier to convert an unused shed into a chicken coop. If you have an outdoor shed that is not used, you may consider converting it into a chicken coop.

The other cost that you will have to meet at the beginning of your project is buying the chickens themselves. Chickens are inexpensive pets, and prices start from as little as $2 depending on the age and breed of chicken you need to go for. If you want to save on the initial buying costs, chicks are generally cheaper than full-grown chickens so you can choose to buy chicks and raise them yourself until they reach egg-laying age.

Once you have your chickens and their coop in place, you will, of course, need to feed them. This means that you will have recurrent costs in terms of buying feed. For egg-laying chickens, the average feed consumption weekly tends to be about 1.5 pounds. This means that with a modest flock, a bag of feed will last you quite some time, and the costs of feeding your chicken will not be high. Chickens are also omnivores, and they tend to eat most things. This means that any leftover food does not need to go to waste since you can feed it to your chickens.

You may also incur additional expenses in terms of veterinary care in case of diseases. Also, plan for costs such as buying feeders and waterers and other miscellaneous items for your coop(s). On the whole, since the chickens you will be keeping will also be providing you with eggs and, for some people, meat as well, the costs and benefits usually tend to balance in favor of the benefits. If you have decided to raise chickens in your backyard, you will need capital investment, but it will not be too high, especially if you just want to keep a small flock of chickens.

Do you have other pets, and can they co-exist with chickens?

Before you bring chickens home, you need to be sure that you have a safe environment for them. If you have other pets, will they be able to share the yard with your chickens? Pets such as cats and dogs may not always be willing to have other animals in their space. They may, therefore, pose a danger to chickens. This is a consideration that you should have in mind, especially if you plan to let your chickens roam free in the yard or live as free-range.

If you have prepared an enclosure for your chickens, make sure it is predator-proof, and yes, this includes ensuring that your other pets will not be able to get at the chickens or harm them in any way. Chickens are susceptible to a lot of predators, and ensuring that they are kept safe will be one of your primary responsibilities. Remember, even friendly dogs or cats can harm chickens, especially if they are still at the chick stage, so always keep them away.

Ultimately, raising chickens is a rewarding hobby, but it is still a responsibility that needs to be taken seriously. Chickens need care and attention to thrive and stay healthy, so before you consider raising backyard chickens, prepare yourself for the responsibility that will come with it.

Chapter 3: Finding the Right Breed for You

The most important decision you will make when you start raising backyard chickens is which breed to keep. For people who have never kept chickens, it may seem like all chickens are pretty much similar. The truth is, however, that there are significant differences between various chicken breeds. This means that for beginners, it is important to understand the key differences between the various chicken breeds and what would make the best match for your needs.

When choosing the best breed, your decision will primarily be based on what you want to keep the chickens for. However, the purpose is only part of what you need to consider. Here are the key factors that should guide your decision on which chicken breed is best for your backyard.

1. Your main purpose for keeping chickens.

2. Your particular climate.

3. The space you have available

Pick a breed based on your main purpose for keeping chickens.

Are you in it for the eggs? While all chicken breeds lay eggs, their production rates and egg size vary from breed to breed. Some breeds are more prolific egg layers, while others are medium egg producers. If your main purpose in raising chickens is to have a steady supply of eggs for your family and perhaps even a surplus for sale, then naturally, you want to go for breeds that produce the most eggs all year round.

The Best Breeds for Egg Production

• Rhode Island Red

This breed is one of the most popular egg-laying breeds in the US and for a good reason. Rhode Island Red chickens can lay approximately 300 eggs annually. This breed lays medium-sized brown eggs. If you are looking for a champion egg layer, Rhode Islands are a safe bet.

In addition to being prolific layers, this breed is pretty low maintenance, which makes it a favorite for people who want to raise backyard chickens. This breed is sturdy, and with good feeding and a comfortable coop, this type of chicken thrives with little attention required on your part. They also tend to have a mild temperament, so they make great pets.

• Plymouth Rock

Plymouth Rock is another prolific egg-laying breed that will do well as a backyard chicken. On average, this breed will lay approximately 300 eggs a year. Another reason to choose this breed is that they adapt well to confinement so they can thrive in small spaces.

This breed is docile and makes great pets since they are not aggressive or territorial. They also have striking black and white plumage that makes for a beautiful flock.

• Australorp

Australorps are great egg layers and can average up to 300 eggs per year. This large breed requires plenty of space due to its large size. This means they will make a great choice if you have a big backyard with lots of space for your chickens.

- **Black Sex Link**

If you want a beautiful bird that still provides you with plenty of eggs all year round, Black Sex Link may be the right breed for you. This breed produces light brown eggs and can average up to 300 eggs per annum.

These prolific layers are a cross between Rhode Island Red and Barred Rock hen. This breed is popular not just for its prowess in egg-laying but also because it is a hardy breed that will not require a lot of maintenance.

- **ISA Brown**

This breed is also a prolific layer that suits people whose primary goal in raising chickens is egg production. ISA Brown chickens can lay up to 300 eggs per year, putting them in league with the best egg-laying breeds. This breed does well in confinement and will adapt well to backyard living. Since they are so docile, this breed also makes great family pets.

Breeds That Produce Blue Eggs

If you fancy an exotic breed that will give you something other than the common brown or white eggs, some chicken breeds lay blue eggs. These breeds include:

- **Araucanas**

This breed tends to be rare, but makes a great backyard chicken and will provide you with blue eggs. This chicken breed is easily recognizable as it lacks a tail head - a feature common in other chicken breeds.

- **Cream Legbars**

Just like Araucanas, Cream Legbars lay blue legs. However, their eggs tend to come in different shades of blue and not just a single uniform blue. This breed is, however, best if you want to raise free-range chickens. This is because it does not adapt well to confinement and will not thrive in containment or small spaces.

- **Ameraucanas**

This is a distinctive breed that stands out with its characteristic beard. These chickens also lay blue eggs and are a great choice for backyard chicken breeders with a taste for the exotic.

Are you looking for a breed that is ideal for meat production?

If you are raising chickens specifically for meat, then you will find that some breeds are better suited for this purpose. Generally, good meat producers tend to be large breeds that typically grow at a much faster rate than egg-laying breeds. This, of course, does not mean that you will not get eggs from the meat-producing breeds. It only means that they are not as prolific in egg production as the egg-laying breeds.

The Best Meat Producing Chicken Breeds

- **Jersey Giant**

True to its name, this breed grows to an impressive size within 20 weeks. It is a favorite breed among people who keep backyard chickens as a source of meat. They will require ample feed to reach their maximum weight.

- **Freedom Rangers**

This is a common meat-producing breed that grows pretty fast. If you do not have the patience to raise the slow-to-mature Jersey Giant breed, you can choose to keep Freedom Rangers. This breed grows to maturity in about 11 weeks. This makes it a popular choice amongst meat breeders. Freedom Rangers are also reputed to have great tasting meat. However, they do require plenty of space to forage and roam in, so go for this breed if you have adequate space in your backyard.

- **Cornish Cross**

Size is an important factor in selecting the best chicken breeds for meat production. This is one of the reasons the large-sized Cornish Cross is a preferred choice for people raising backyard chickens for meat. This breed grows fast and will reach full maturity in about six weeks. It is renowned for its large thighs and ample breasts, which are good qualities to have in meat-producing breeds.

- **Bresse**

Bresse is not the fastest-growing meat producing chicken breed, but this breed is popular among people who are looking for quality meat. This breed weighs in at approximately 7 lbs. So, it may not be as large as the other meat-producing breeds but is a great choice if you want a breed that is not too large but still adequate for meat production.

Do You Want a Dual-Purpose Breed?

For some people, the ideal chicken breed is one that can be a source of both eggs and good quality meat. If these sounds like just what you need in your backyard chickens, these are the breeds you need to consider.

- **Marans**

Marans are a chicken breed that can serve as both a medium egg-producer and as a source of meat. This breed is available in a variety of colors, including copper blue, black-tailed buff, and golden cuckoo, among others. These chickens are a hardy breed that does not require a lot of care and maintenance. They make great backyard chickens since they adapt well to confinement and are generally mild-tempered. This breed lays dark brown or chocolate-colored eggs.

- **Sussex**

This breed is popular across the globe and is one of the best breeds if you want to raise dual-purpose backyard chickens. They do well as free-range chickens and are good at foraging for food. If you want a chicken that is child friendly, the docile Sussex meets this criterion and makes a great family pet.

- **Wyandotte**

This breed comes in a variety of colors and is often raised as a show bird. However, if you want a breed that will serve as both a source of eggs and meat, the Wyandotte does both well. It also makes a great backyard chicken since it adapts well to confinement and is naturally mild-mannered.

- **Turken**

This breed is also commonly referred to as naked-neck chicken owing to the fact that it does not have any feathers on its neck. This breed tends to be a hardy, low-maintenance chicken that is ideal both as a source of fresh eggs and meat. This breed is generally not fussy and is mild-mannered enough to make a docile family pet.

Which Breeds Make the Best Pets?

Some people keep chickens simply because they want a pet and a pleasurable hobby. If this sounds like you, then you need to know which breeds are ideal as pets. When you are looking for a breed that will make a good family pet, you need to consider the temperament of each specific breed. Some breeds can get quite broody and may attack, especially when they have small chicks.

If you are looking for a docile breed that will be safe even around children, these are the breeds to go for:

- **Plymouth Rocks**

This is one of the most docile chicken breeds and is ideal if you want a family pet; as an added plus this breed is a prolific egg-layer, so you get the best of both worlds if you choose to have this breed in your flock of backyard chickens.

- **Buff Orpington**

Orpingtons make great pets due to their docile and easy-going nature. They are friendly and make calm pets that even children can help to care for. This breed is also dual-purpose, which means it will provide you with a decent supply of eggs and meat if required.

- **Australorp**

What can be better than a friendly pet that also provides a steady supply of eggs for you and your family? If this sounds like just what you need, then the Australorp breed is an ideal choice for your backyard flock. These mild-tempered chickens are friendly and curious, and they do well around people.

- Cochin

Though quite big in terms of size, the Cochins are gentle giants that are usually calm and friendly. This is a bird that enjoys a cuddle. This fluffy bird likes to have some lap time and easily bonds with its owner. Since this breed is also a medium egg-layer, you get a friendly pet as well as a source of fresh eggs if you make this breed part of your backyard chicken flock.

While chickens, on the whole, cannot be considered aggressive pets, there are some breeds that can get quite broody. The Silver Laced Serama is often touted as the most aggressive chicken breed, so you may want to steer clear of this breed if what you are looking for is a friendly family pet.

Bantam Chicken Breeds

Bantam chickens differ from regular chicken in one major aspect: size. A Bantam chicken will typically be roughly a quarter of the size of regular chicken breeds. This small breed of chickens is a good egg-layer and comes with the added advantage of consuming less feed than the other chicken breeds.

If you have a small backyard, Bantam chickens make a great choice since their small size means they can be adequately housed and free-ranged in smaller spaces. Here are some of the reasons you may want to make Bantam Chickens your choice when choosing a breed to raise in your backyard.

I. They are good egg-layers. Each chicken produces about 4-5 eggs a week.

II. They make great pets due to their docile nature and small stature.

III. They require less feed than other breeds making their maintenance cost lower than that of normal-sized chicken breeds.

IV. These little birds are adorably cute and will make for a beautiful flock. Some people even raise them as show chickens.

Consider the best breed for your particular climate.

Different chicken breeds will thrive in different climates, depending on their natural resilience and the adaptabilities they have developed over time. If you are a beginner, it is best to go for breeds that are suited to the climate in your area. This will help to minimize the risk of diseases for your chickens and boost their overall wellbeing.

Best Breeds for Cold Climates

For cold areas, go for breeds that are suited to the cold weather. These breeds will have lots of feathers on their bodies to help keep them warm. Most of them also tend to have feathered legs, which again helps to keep the bird warm. As a natural adaptation to cold climates, breeds that do well in cold conditions will have small combs. This helps them to avoid frostbite.

If you live in a cold climate, these are the breeds you should aim to make part of your backyard flock.

● **Rhode Island Reds**

This prolific egg-laying breed is well suited to cold weather and will do well in cold climates. Their plush feathers effectively keep this breed of chickens well insulated from the elements.

● **Australorps**

Just like Rhode Island Reds, Australorps have heavy plumage, which helps them stay warm even in cold conditions. When you go for birds that are adapted to colder climates, you are sure that they will thrive in your backyard and you will not have to deal with constant ill-health or even birds dying due to adverse weather conditions.

- **Brahma**

This breed has the characteristic feathered feet that make some chicken breeds better suited to cold climates. This big chicken breed is docile and very friendly and therefore makes great family pets. It will also keep you supplied with eggs, although it is better known as a meat-producing breed.

If you are in a cold climate and want a hardy breed that is built to withstand cold conditions, Brahma chickens are a good choice for your flock.

- **Dominique**

This chicken breed is small in stature but is equipped with enough plumage to keep it warm in colder climates. In fact, this breed does not tolerate heat well and is therefore ideal for you if you live in a cold region and need a chicken flock that is suited to that particular climate.

- **Ameraucanas**

Ameraucanas are famous for their blue eggs, but this particular breed also thrives in cold conditions. Although it is not the most prolific egg-layer, you will still get a decent supply of eggs from this breed.

Best Breeds for Hot Climates

If you leave in a hot climate, then similarly, you will need chicken breeds that are suited to that particular climate and can survive the high temperatures. In terms of raising chickens, areas that are classified as hot are those that average 89.6 degrees Fahrenheit or higher. For breeds to do well in such hot weather, they need to have natural adaptations that reduce the effect of the heat on the chicken. These adaptations include lighter plumage, lighter colors that do not absorb as much heat, and smaller bodies.

For hot climates, these are the breeds that will thrive and do well.

• Plymouth Rock

We have already covered this particular breed under the best breeds for egg-laying as well as its suitability as a family pet. These attributes make it one of the most popular backyard chicken breeds for hot areas. This hardy breed is adaptable and will do well in both cold and hot conditions, making it one of the most versatile breeds you can have in your flock.

• Golden Buff

This breed is hardy and adapts well to hot climates. It also does well in cold climates so you can keep it regardless of what kind of climate you live in.

• Leghorn

For hot climates, the Leghorn breed stands out due to its sturdy and resilient nature. These birds are good egg-layers and are recommended for people who want a breed that is good for egg-laying and thrives in hot climates.

• Fayoumi

Fayoumis are striking birds that are hardy enough to thrive in extreme heat conditions. They are well adapted to hot climates and would suit you if you want a striking flock of exhibition chickens.

Pick breeds based on the space you have available.

Large breeds will naturally require more space, and therefore you should be aware of the size of the breed you are buying. Often, if you are buying chicks you may not be able to estimate the potential size of the full-grown chicken. While space will only become a pressing issue if you want to keep large flocks, it is still best to ensure that you have adequate space for the particular breed you want to purchase.

When chickens get overcrowded in small spaces, the risk of infectious diseases and parasites spreading amongst them increases significantly. This can end up being costly, and it is therefore best to avoid the situation altogether. Most breeds that are best for meat production tend to be larger than egg-laying breeds, which means that if you want to raise chickens for meat purposes, you will probably need more space.

Large chicken breeds that require plenty of space include breeds such as Jersey Giant, Cochin, Brahma, Cornish, Orpington, Rhode Island Red, and New Hampshire. Though these breeds are large, it is still possible to breed them on a small yard as long as you stick to a modest flock to allow each bird enough living space.

It is also important to note that apart from more space, raising large chickens is pretty much the same as raising smaller and medium-sized breeds. Some people mistakenly think that larger breeds are more aggressive. This is not true in any way as a chicken's temperament is not connected to its size. In fact, most of the more friendly and docile breeds tend to be large breeds like the Jersey Giant, Rhode Island Reds, Cochin, and Plymouth Rocks.

Ultimately, whatever breed of chicken you choose for your backyard flock, you will need to care and nurture for them well to enable them to thrive. Taking good care of your backyard chickens by ensuring that they are well-fed, housed adequately, and have a clean, safe environment will be the main factors in determining whether or not you get the best out of your backyard chicken flock.

Chapter 4: Getting Set Up and Selecting a Coop

Before you bring the chickens home, you will need to set up their living area. This means having a coop to shelter your chickens and a chicken run for them to forage in, as well as all the other materials required for the proper care of chickens. If this is your first time raising chickens, then you probably need to start your setup from scratch. This means identifying where you want your chicken coop to be, how large or small it is going to be, and whether or not you will free-range your chickens.

Chickens, just like any other pet, have basic requirements that need to be met. These requirements are the factors that will guide you on how to prepare your backyard for your chickens before you bring them home. You want to avoid a situation where you bring your flock home only to realize that you are missing something essential.

These are the basic requirements that are necessary for your chickens to be healthy and happy.

- A well-constructed shelter to house the chickens, which is the chicken coop
  - Food and water

- Enough space to move around
- A chicken run or forage area for them to dig, scratch, etc.
- A safe nesting place for broody hens

Choosing the Right Location for the Chicken Coop

When it comes to preparing to raise backyard chickens, the first thing you need to figure out is the coop, which is basically where your chickens will be sheltered. The chicken coop poses several questions that you need to answer before you can identify the right coop for your particular needs; things like how big a flock you want, whether to build or buy the coop, the appropriate size, and whether or not you want a stationary coop. However, even before you get to all that, you must first figure out where to position the chicken coop in your yard.

Location is an important consideration when it comes to providing appropriate shelter for your chickens. Where you place your chicken coop entails factors like how much sun and shade your chickens will get, wind exposure, safety, convenience, and a host of other key factors. Besides, chicken coops can get quite smelly, and they also tend to attract insects. This means that if you position the coop too close to your house, you may have to contend with unpleasant odor and bugs.

To ensure you get the location of your coop right, keep the following considerations in mind as you pick the ideal spot for your chicken coop.

1. Distance from the Chicken Coop to Your House

The general rule is to ensure that you do not place the chicken coop right next to your house. Since chicken poop tends to have a strong odor, this smell can easily become a nuisance if the chicken coop is too close to your house.

You will also find that chickens tend to attract insects and bugs, which you, of course, do not want to become a permanent fixture in your home. When choosing the best location for your coop, identify a spot that is not directly next to your house but is still close enough to access conveniently.

Having the coop not too far away from your house means you can easily check on them when needed. Since you also need to feed, water, and collect eggs from the coop, this means you will make trips to the coop daily, so having it close by will make your chores a lot easier.

In case you need to connect electricity to your coop for heat or any other reason, it will be easier if it is not too far away from your house. This also goes for amenities such as water for cleaning and watering your chickens. In a nutshell, find a location for your coop that is not directly adjacent to your house but is also not too far off.

2. Find a Location with a Level Ground

It is important to ensure that the chicken coop is situated on level ground. This will help to ensure that the structure is stable and durable. You can clear a level patch for your chicken coop. Remember that the area will also need to have good drainage since you do not want your coop getting submerged in water during the wetter months.

If you live in an extremely wet area putting in a concrete foundation will make your chicken coop structure more stable and durable. Some coops are constructed with floating floors, which basically means that the floor is suspended above the ground on concrete blocks to create a level surface.

3. Your Coop Should Have Enough Foraging Area Around It

Chickens love digging around for bugs, foraging, and looking for food on the ground. You therefore need to ensure that your coop has sufficient forage area around it. The size of the forage area will, of course, depend on the size of your flock, but having at least 8 square feet per chicken is best. An ideal forage area can be a combination of a grassy patch and dirt.

When you do not allow enough space for chickens to roam, they become more prone to infections and ill health. If you are confining your chickens to a run, make sure it has sufficient space for them to forage based on the size of your flock. For free-range chickens that are not confined to a run, you will still need to make sure your yard has enough forage area for the number of chickens you intend to keep. For free-range chickens, the recommended forage area is approximately 250 square feet per bird.

4. Your Coop Should be in an Area That is Not too Windy

You want your chickens to be nice and warm in their chicken coop, especially during the colder months. This means that when picking the location for your coop, consider whether the area has a windbreak. Positioning your coop in an area with some windbreaks such as trees or a tall structure will ensure that the temperatures in the coop are not adversely affected by windy conditions.

5. Pick an Area That Gets Some Sun but Also has Some Shade

Your coop should be in an area that gets some sun. It should also have some shady areas where your chickens can seek respite from the sun in hotter months; chickens thrive in a setup where they can enjoy both some sun as well as some shaded areas when it gets too hot. If you can find a sunny spot with a few trees that can offer some shading for the coop and forage area, that will be an ideal location.

How to Choose the Right Chicken Coop

Once you have figured out the perfect location for your chicken coop, the next step, of course, is to identify the best chicken coop for your needs. There are plenty of options when it comes to chicken coops. The variations in size, shape, and design mean that there is a coop available to suit different requirements. However, you do need to know what makes a good chicken coop before you settle on a particular design.

These are the main considerations to keep in mind when choosing a chicken coop.

I. Size

II. Internal structures; roosting bars, and nesting boxes

III. Ventilation

IV. Safety

Size

The first consideration when choosing a chicken coop is to ensure that the coop is the right size for your flock. Depending on how many birds you want to keep, and the size of the breed you have chosen, the amount of space you need in the chicken coop will vary.

You must ensure that your coop is the right size for the flock you intend to keep. While you can allow more square footage than the recommended guidelines, do not go overboard. A coop that is too large may be colder and require extra heating to keep your chickens warm.

For large breeds, such as Jersey Giants, Plymouth Rock, or Rhode Island Reds, you will need a minimum of 4 square feet per bird in the coop. This is the minimum space per bird, and you can always have a higher allowance of space for your chickens.

For medium-sized breeds such as the Leghorn, you should have a space allowance of at least 3 square feet per bird. Again, this is the minimum space, so you can always have a higher space allowance for your chickens.

Smaller breeds such as Bantams do not require a lot of space. An average of 2 square feet per chicken should be enough if your flock is made up of small breed chickens.

When your coop is too small for your flock, it can cause the following problems:

• Poor chicken health due to high ammonia levels in the coop from accumulated chicken manure.

• Poor egg production due to cramped conditions in the coop.

- Bullying and aggressive behavior among the flock as they each fight for space.

Internal Coop Structures: Roosting Bars and Nesting Boxes

Once you have calculated the square footage of the coop that will be adequate to house your flock of chickens, you then need to consider the appropriate size of the structures inside the chicken coop.

One of the most important structures inside your chicken coop will be the roosting bars. Chickens do not sleep on the ground. They need roosting bars raised from the surface of the coop. The roosting bars need to be higher than the nesting boxes in the coop.

The roosting bar needs to provide adequate room for each bird inside the coop to avoid overcrowding. Your roosting bar should provide approximately 8 inches of roosting space per chicken. During colder months, chickens tend to draw closer to each other, so do not make the roosting bars too big.

Nesting boxes provide a private space for your chickens to lay eggs and for broody hens. While you do not need to have too many of them, ensure that you have at least one for every three chickens in your flock. This means that for a flock of 12 chickens, four nesting boxes will be sufficient. Having enough nesting boxes ensures that your chickens have a safe place to lay their eggs.

The Chicken Run

Apart from the inside of the coop, you will also need to ensure that the outdoor space for your chickens, or the "chicken run", is the appropriate size. Remember that when your chickens are not in the coop, they will be foraging outdoors in the run, so it is an important extension of the chicken coop. The recommended square footage for the chicken run is at least 8 square feet per bird.

If you plan to free-range your chickens, you may not need an outdoor run. However, even for free-range chickens, having a confinement area may come in handy when you need to pen the chickens up for a while for one reason or another.

Ventilation

The chicken coop needs to be properly ventilated, and it should allow sufficient air circulation. This means that your coop should have enough air vents to let air in and out of the coop. When ventilation is poor, the build-up of ammonia in the coop from the chicken poop can cause ill health to your chickens. Make sure the air vents are well secured with chicken wire to prevent predators and rodents from getting into the chicken coop.

Safety

Unless you want to wake up one day and find that a wily fox or raccoon has fed on your chickens, you will need to make safety a priority when choosing a chicken coop. Chickens have plenty of predators, ranging from the average domestic dog or cat to possums, foxes, and raccoons. This means that your chicken coops should offer adequate protection for your chickens.

A secure chicken coop should not have gaps and spaces that can allow rats or snakes to enter the coop. Even the ventilation spaces should be well covered with chicken wire to keep rodents and predators out. Also, ensure that the doors are well secured with childproof locks that stay locked. Keep your chickens locked into the coops every night.

It is always advisable to keep your chicken feeds in a separate area and not in the chicken coop. This is because some predators will be attracted to the coop because of the chicken feed. The same should also go for chicken eggs. Do not get into a habit of leaving them uncollected in the coop for days, as they will attract predators to the coop. Some predators are more interested in the chicken feed and the chicken eggs than the chickens themselves, so if you can avoid putting feed in the coop, you will deter such predators.

Your chicken coop should also have a secure roof. This will serve to keep predators out and also ensure that your chicken coop is kept moisture-proof, especially during wet weather. Chickens do not like to be rained on, so providing them with a well-covered coop is important.

Safety precautions will also be needed for the chicken run since your chickens will also be spending time outdoors. Your run should be enclosed using fencing materials that have very small openings such as chicken wire, welded wire mesh, or electric netting. This will ensure that even when your chickens are outdoors, they are well protected. In areas with flying predators such as hawks and owls, overhead fencing can be used to make your chicken run more secure.

If you are looking to raise your chickens as free-range, you will need to make sure that your yard is well secured so that your chickens can forage safely. This means making sure that your yard fencing is intact and high enough to keep out predators. You can also bury welded-wire mesh or any other small mesh fencing material to deter digging predators that tend to dig holes to access the chickens in the run.

Ensuring your Coop is Chicken-Ready

Feeders and Waterers

Once you have found the right location for your coop and found the best coop for your needs, the next step is to make sure that your coop is equipped to shelter the chickens. Naturally, you will need to feed your chickens, so you will need to get chicken feeders and waterers for the coop.

Chicken feeders come in a variety of shapes and sizes. The best feeder for your chickens will depend on the size of your flock. There are also smaller feeders that are designed to be used specifically for chicks, so make sure you buy the appropriate feeder.

The most common type of chicken feeder is the dispenser feeder. This feed dispenser gradually releases feed as it is consumed. These types of feeders can be strung up to keep dirt and debris out of the feed and also to discourage bugs and rodents.

Some people choose to go for automatic feeders that only need to be refilled occasionally. This can be a good choice if you want a feeder that does not need to be refilled every other day. However, you do not necessarily need an expensive feeder to get the job done. Even a homemade chicken feeder from recycled materials such as plastic containers will serve the same purpose as a store-bought feeder. You can easily make a homemade chicken feeder using a bucket or any other type of plastic container.

It is always advisable to keep your feeders inside the coop to protect the feed from the elements. However, you can still choose to have a feeder in the chicken run provided you position it where it is safe from rain, dirt, and debris. To make sure all your chickens have access to a feeder, ensure that you have enough feeders to accommodate the entire flock. If your feeders are too few, the smaller chickens may get bullied during feeding time. Have at least one feeder for every ten birds.

Apart from a chicken feeder, you will also need a waterer or drinker for your chickens. Chickens need to stay hydrated, and they must have access to clean drinking water. Waterers, just like feeders, come in a variety of shapes, sizes, and designs. Consider the size of your flock when selecting a waterer. Some waterers need to be wall-mounted, so only go for this type if the coop you have can accommodate a wall-mounted waterer.

The most common types of waterers are gravity fed waterers, automatic waterers, and container waterers. Gravity-fed waterers are popular because they are easy to use and, therefore, the most convenient. Automatic waterers are ideal if you are short on space since they come with cup or nipple attachments for the chickens to drink from. You may need to train your chickens on how to drink from the automatic waterer, but most tend to get the hang of it pretty fast.

Most waterers are made out of either steel or plastic. Both of these materials are durable. However, steel waterers can heat up considerably in hot weather and may not be ideal if you live in a hot climate. Plastic waterers are generally cheaper than the steel ones, so they are a good option if you are on a budget.

Keeping your waterer inside the coop can lead to wet bedding. To avoid this situation, most people choose to put their waterers outside the coop in the chicken run. You can also reduce the risk of spillage and leaks by not overfilling your waterers. Just like with a chicken feeder, ensure that you have enough waterers for the size of your flock.

Chicken Coop Beddings

Chickens poop a lot, and having bedding in your coop will help to keep the coop clean and odor-free. The bedding that you will use on the floor of your chicken coop will serve as litter to aid in controlling odors and moisture in the coop. Bedding will also provide insulation for the coop, so it is good practice to put bedding in your coop before you bring your chickens home.

The best type of bedding for a chicken coop is an absorbent material that will aid in keeping the coop floor dry. A wet coop floor can lead to diseases and may cause lesions on your chickens' feet. So, when picking the best bedding for your coop bear in mind that you need to have a material that will absorb and release moisture quickly. Common bedding materials include wood shavings, straw, hay, and grass clippings.

Wood shavings make one of the best options for chicken coop beddings. They are absorbent but do not retain moisture for long, so your chicken coop floor stays dry. You can put newspapers under the wood shavings for easier cleanup but do not use newspapers alone as bedding in your chicken coop.

For the nesting boxes, you can use straw and hay to cushion your chickens when they are laying eggs. This will also ensure your eggs do not break. Alternatively, you can also use the same wood shavings you have used on the rest of the coop floor in your nesting boxes.

Ultimately the whole setup of the chicken coop and chicken run should be centered on providing your chickens with a clean, safe, and comfortable living environment.

Chapter 5: Building a Chicken Coop

Once you have set up your yard and decided on the kind of chicken coop that you will need for your backyard chickens, you have three options; buy a pre-made chicken coop, repurpose an existing structure, or build one yourself. Depending on your budget, the kind of design you want, and how handy you are with DIY projects, the choice will be easy to make.

Some people, especially those with big pasture areas, opt to have portable chicken coops. This type of coop can be moved from one section of your land to another, in effect allowing the chickens to forage on different sections of pasture. These types of coops work well if you have plenty of space and pastureland. However, for small backyards, especially in towns and cities, a stationary coop may work better since it does not require repeated moving.

If you do not have the time or the know-how to build your own chicken coop, you can always go for a pre-made chicken coop. These are widely available in online stores such as Amazon, as well as in pet shops and even grocery stores like Walmart. Pre-made coops come in various sizes, designs, and price points so you can go for one that falls within your budget while meeting all the requirements required for the number and type of chickens you plan to keep.

Repurposing an existing structure is also a simple way to create a chicken coop. If you have a shed that is no longer in use, you can repurpose it to serve as your chicken shed. All you will need to do is outfit it for chickens by adding nesting boxes, roosting bars, and some bedding on the floor. Ultimately, this beats building a chicken coop from scratch and is, of course, much cheaper than buying a pre-made coop.

The third option is building your own chicken coop. This gives you the freedom to make a customized chicken coop that suits your backyard and your needs perfectly. You have a choice of getting designs done by a professional or doing them yourself. There are also plenty of online resources that offer free chicken coop plans. Before you settle on a design, always check if there are any town ordinances or neighborhood regulations that need to be met.

If you choose to build your own chicken coop, there are many ways to do it depending on the kind of chicken coop you have in mind. You can choose to have it done by a professional if you are short on time or you do not have access to the materials required. Alternatively, if you want a simple and quick way to do it, you can follow our simple step by step guide for building a 24-square-foot coop that can hold between 6 and 8 chickens.

Materials and Tools Required

- Wood
- Hammer
- Saw
- Measuring Tape
- Pencil
- Screwdriver
- Electric Stand Saw
- Extension cords

- Spirit Level
- Sandpaper
- Paint Brush

While coops can be built using a variety of different materials, wood is the easiest material to use. Wood is also good for insulation, and it makes solid and stable structures that are durable.

Step by Step Coop Building Process

Once you have your supplies ready, you can start building.

1. Start by Building the Floor of Your Coop

- Start with a piece of plywood cut to 4 feet wide and 6 feet long.

- The plywood should be at least half an inch in thickness. This will ensure that your floor is sturdy.

- To make your floor frame, you will need battens. Ideally, these should be 2 x 4s. Screw the 2 x 4s around the borders of the plywood. Also, screw another 2 x 4 across the middle of your plywood floor.

2. Build the Solid Wall Next

- The solid wall of your coop is the one that will not have a window. Take a ½" (or thicker) piece of plywood 6 feet long. You will need 2 x 2s for this wall. Secure the 2 x 2s to the bottom of the vertical edges of your plywood. The 2 x 2s should stop 4 inches above the bottom of the plywood.

- Once you have screwed on the 2 x 2 s, you can now secure the wall to the floor you built in step 1. Take your solid wall and place it on the floor in such a way that the 4 inches that you left cover the 2 x 4s on the underside of the floor. Once you have positioned the wall, screw it in place. Your screws should be 1½" in order to secure the wall firmly to the floor.

3. The Next Step is the Front Panel

- Attach a four-foot length of ½" (or thicker) plywood to the floor and the solid wall you have already built. First, screw the piece of plywood to the two 2x4s at the bottom of your coop; then secure the plywood to the solid wall by screwing it on to the two 2x2s on the solid wall.

- Once the plywood is secured to the coop, it is time to cut the door.

- The door opening should be 2-to-3 feet in width. The height can vary, provided you leave a minimum of about 6 inches between the edge of your door and the bottom part of your plywood panel. The same 6-inch allowance should be left between the edge of the door and the top of the plywood panel.

- Once you have marked out the measurements for the door opening, cut it out with your saw. Make the cut as smooth as possible.

- You will want to reinforce the top of the door opening by using a 20-inch piece of wood. Attach this piece of wood to the top using screws and some construction glue.

4. Let's Build the Back Wall

- Just like the front panel, for the back wall, you will also need a piece of plywood that is 4 feet in length and at least ½ an inch in thickness.

- Secure the piece of plywood to the back of your coop by screwing it to the 2x4s on the underside and then screwing it to the 2 x 2s on the solid wall of the coop.

- Once the back wall is secured to the coop, you can now measure out the door opening for this wall. Using the same measurements that you used for the opening on the front panel, cut out the opening just as you did for the front panel.

- Finally, reinforce the top of your door opening with a piece of wood just as you did to the opening on the front panel.

5. Build the Last Wall

- Cut two pieces of ½" (or thicker) plywood to a length of 2 feet. Next, cut a piece of plywood 5 feet long. The width of this piece should be half the height of your coop.

● Once you have these three pieces of plywood, you can start securing them to the coop to build the last wall. Start with the 2-foot-long pieces of plywood. Secure a 2 x 2 to one of the vertical edges of the plywood. The two 2x2s should stop at least 4 inches above the bottom of the plywood.

● Take the second 2-foot length of plywood and also attach a 2x2 to one of the vertical edges of the plywood. The 2x2s should leave a 4-inch allowance to the bottom of the plywood.

● Now take one of these plywood panels and attach it to the front of the coop. Once that is done, take the second panel and secure it to the back of the coop.

● Now take the 5-foot-long piece of plywood and secure it between the two panels you have just attached.

● The edge of the 5-foot plywood should line up with the tops of the other two panels.

● The next step is to take a piece of wood that is the same vertical length as the middle piece. Screw this piece of wood to the joint where the middle panel connects to the side panel. Do the same for the second joint where the middle panel connects to the other side panel. This way, you have two pieces of wood, reinforcing the two joints between the middle panel and the other two panels.

6. Constructing the Roof

■ For the roof, you will start with the gables. These are the triangular structures that will be placed on top of the walls of the coop to support the roof.

■ To fit properly on the walls, you need to make your gables 4 feet long. Make sure that the pitch you create for both gables is the same so that your roof seats evenly on the coop.

■ The gables will go on the top of the front and back walls. Take the first gable and secure it to the inside of the front wall. Use screws and some construction glue to attach it securely.

- The second gable should be attached to the inside of the back wall. Make sure that the attachment is secure.

- Once the gables are attached, you need to build support for the middle of the roof, which is the truss.

- The angle of your truss should be the same as the one on your gables.

- To make sure you get the right angle, take two pieces of 2 x 2s and clamp them to the edges of one of the gables. The 2 x 2s should be longer than the edge of the gable by about 3 inches.

- You will need a crosstie to reinforce your truss. This crosstie needs to be of similar length to the gables.

- Attach this crosstie to the 2 x 2s with screws. Once it is attached, you can remove the truss from the gable by removing the clamp you had used to attach it to the gable.

- Now place the truss in the middle of the coop.

- Make marks where the 2 x 2s of the truss intersect with the side walls. These marks represent where you will make the notches on your truss.

- Once the notches are made, you can now place the truss on top of the side walls. It should be at the center of the two side walls.

- Now that the roof supports are in place, you need to make the actual roof.

- Using two pieces of plywood, make a roof by joining one 40-inch piece of plywood with an 84-inch piece of plywood. The joints should be along the longer 84-inch sides. You can easily join these two pieces using hinges.

- The roof is now ready to go on top of the coop. It will have overhangs on both sides of the coop.

- You will need to attach two pieces of 2 x 2s to the bottom edge of the front and back overhangs of the roof.

- Once the trim is in place, the final part of constructing the roof is to secure it to the gables on each side and the truss in the middle.

- You can then make your roof moisture-proof by covering it with tar paper or galvanized roofing.

7. Building the Coop Doors

• Now that the walls are finished, it is time to build the doors.

• Take a medium density fiberboard and cut it to the same length as your door opening and half the width of the door opening.

• Construct your door frame using 2x2 pieces of wood. Fasten these pieces on the four sides of your door opening.

• Once the frame is in place, you can now screw in the hinges. Use two hinges for each door.

• Once the hinges are attached, you can now fix the doors to the frame.

• You will then construct the back doors using this same process.

• Once all the doors have been attached to the coop, you now need to put in locks so that the coop is lockable. Get secure locks for your doors that will effectively keep predators out of your coop.

8. Make Legs for Your Coop if You Want it Raised

If you want your coop to be raised, you will need to attach four pieces of 2 x 4s to the underside of the chicken coop. You can secure these legs to the 2 x 4s on the bottom of your coop.

If your coop is raised, you will need a ladder to make it accessible to the chickens. For your ladder, attach 2 x 2s to 2x4s to the required length. Take your ladder and then secure it in place with some hinges.

9. Roosts and Nest Boxes

The interior of your coop will need two essential structures. These are roosting bars and nest boxes. Roosting bars are basically raised bars where your chickens will perch and sleep at night. Chickens do not sleep on the ground, so they need roosting bars that are elevated off the floor of the coop.

For your roosting bars, allow at least eight inches of space per chicken. You can put in multiple roosts depending on the size of your flock. For your roosts, you can use sturdy pieces of wood secured to the coop wall at an angle or even a short ladder propped up at an angle. The roosting bars need to be at least two feet off the coop floor.

The other essential structures to have in your chicken coop are nest boxes. These boxes provide your chickens with a private area to lay eggs. Nest boxes will also be used by the brooding hens when they want to hatch eggs. On average, you will need one nesting box for every four birds.

You can construct one-foot square wooden boxes and use them as nesting boxes in your chicken coop. Alternatively, you can easily repurpose old milk crates and use them as nesting boxes. Whichever type of nesting boxes you choose, simply secure them to your coop walls, and you are all set.

10. Coop Bedding

The last step to prepare your coop and make it ready for your chickens is the bedding. Bedding is an absorbent material that is put on the floor of the chicken coop. The bedding helps to keep the coop floor dry by absorbing moisture from the chicken coop. It also absorbs odor from chicken manure, and this helps to prevent ammonia build up in the coop. Another advantage of having bedding in your coop floor is that it makes clean up easier. It also provides additional insulation for the coop, helping to keep your chickens nice and warm, especially during the colder months.

Wood shavings make great bedding since they absorb and release moisture. Other materials that can be used as bedding are straw, hay, sand, and grass clippings. Be sure to also cushion your nesting boxes with some straw and hay.

Tips for Your Chicken Run

If you do not intend to raise free-range chickens, then you will need to build a chicken run to keep your chickens confined when they are outdoors. Chickens need a foraging area to dig and forage outdoors, so if you do not want them roaming all over your yard or garden, you will need to confine them to a particular section in your yard.

Typically, the chicken run should be adjacent to your coop so that your chickens can get in and out of the coop from the run. Basically, once you have decided the appropriate area that is sufficient for the size of flock you are planning to keep (work with a minimum of 5 square feet per chicken), you can fence this area off to make your chicken run.

To make sure that your run will keep your chickens safe from predators and also keep them from wandering, you should use chicken wire, welded wire mesh, or electric netting as your fencing material. When you use small mesh fencing materials, you effectively keep your chickens safe from smaller predators that can reach through a larger mesh or even jump through it.

If your area has plenty of flying predators such as hawks and owls, you can choose to cover your chicken run with chicken wire or any other small mesh fencing wire. Remember that you want your chickens to be safe in their run. You also do not want to have to keep checking on them constantly to see if they are safe. This means that taking all the safety precautions necessary when setting up your coop and chicken run will save you time in the long run and save you plenty of predator trouble down the road.

Chapter 6: Tips on Buying Chickens

When your coop is all done and you are eager to bring your feathered friends home, then it is time to buy your chickens. Of course, by now you already have an idea of the number of chickens you want, and what kind of breed is best for you. With this in mind, you can now start shopping for the right birds for your backyard.

Where Can You Buy Chickens?

There are plenty of options for people who want to buy poultry. Whether you are looking for day-old chicks, pullets, or mature chickens, there are plenty of hatcheries, feed stores, poultry associations, and breeders where you can buy your flock. It is important to look for reputable hatcheries and breeders so that you can be sure that you are getting birds that are in good health.

If you are not aware of any breeders in your area, check with any farm stores in your area. Usually, they will have information on local breeders or hatcheries. Alternatively, most breeders have some sort of online footprint, so you may get some leads to reputable breeders in your area by checking online sites and social media platforms. If you want a specific breed, you can also check on social media for breeder groups that specialize in the particular breed that you have in mind.

If you choose to go with an online breeder or hatchery, always check consumer reviews and feedback to ascertain that the supplier is credible. Some people prefer to buy their chickens from other farms since they can see the kind of environment the chickens have been bred in. As much as possible, when buying chicks, pick a hatchery or breeder that is closer to you to minimize the amount of time your chicks have to spend in transit.

Farm stores are easily accessible to most people and are a popular place from which to buy chicks. However, when you buy from a farm store, you will not know whether the chicks are male or female, so you may end up with roosters that you did not want. You may also not be able to get information on whether the chicks have been vaccinated or not.

The other alternative is buying from a hatchery. Hatcheries usually have a variety of breeds available and tend to be cheaper than breeders. Hatcheries tend to specialize in utility birds and may not really be a good source if you want heritage chickens. For rarer breeds, breeders are usually a better source. Breeders tend to specialize in particular breeds. They can be a bit pricey compared to hatcheries, but on the upside, you can get even heritage chickens from breeders.

If there are no hatcheries or breeders near you, some hatcheries offer shipping options across the country and will get your birds to you wherever you are. These include:

- **My Pet Chicken**

This hatchery is great for beginners looking to start with a small flock. You can order as few as three chicks. They also carry different breeds, so you have a wide range of options to choose from. As a plus, they sell other accessories and chicken equipment that may come in handy for beginners such as coops, chicken fencing, feeders, and more.

- **Cackle Hatchery**

This hatchery in Missouri has all types of chickens on offer. From layers to meat breeds and dual-purpose birds, you get most types of breeds from this hatchery. Since they allow even small orders, you are not required to buy in bulk, so this is also a good place for the urban dweller who wants to keep a modest flock.

- **Murray McMurray**

This hatchery is located in Iowa and has a wide range of chicken breeds to choose from. They also have various equipment and accessories for chickens, so this can be your one-stop-shop for your chicken needs.

- **Freedom Ranger Hatchery**

Freedom Ranger is ideal for people looking for free-range chickens. This hatchery uses eco-friendly farming methods, and it is well known for its organic free-range poultry.

- **Meyer Hatchery**

This hatchery has over 160 poultry breeds for buyers to choose from. They offer gender-guarantees, so for people who want to buy chicks that are strictly female or male, this is a great hatchery to buy from.

- **Ideal Hatchery**

This Texas Hatchery guarantees 100% live delivery to its clients. They also have plenty of breeds to choose from. However, they do have a minimum order requirement, so they may not be the best option if you want a small number of chickens.

- **Stromberg's Chicks**

This hatchery has locations in five states, including Minnesota, California, Texas, Pennsylvania, and Florida. They have an impressive selection of chicken breeds to choose from, and if you are looking for accessories and equipment as well, they have plenty of those, too.

Information You Need from Your Breeder

Once you have identified the breeder or hatchery you will be buying your chickens from, here are some basic questions you need to ask the breeder.

1. Find Out What Breeds are Available

Most breeders will specialize in a few select breeds, so you need to know what type of breeds they have. You can then decide whether the breeds they have are what you require or move on to a different breeder.

2. Find Out if They Have Sexed Birds

If you are buying chicks, it is not physically possible to tell whether the chick is male or female at that young age. Sexed birds are those that have been checked by the breeder and ascertained to be either male or female. This identification is important, especially if you live in an area where roosters are not allowed.

When you buy sexed birds, you know exactly what you are getting, and there will be no nasty surprises later when one of your chicks turns out to be a noisy rooster.

3. Find Out if the Breeder is Certified by NPIP

Breeders who are certified by the National Poultry Improvement Plan are those that have assented to having their chickens checked for diseases. If the breeder is certified, you will have some assurance that the birds you are getting are in good health.

4. Find Out if the Chickens Have Had Any Vaccinations

When buying chicks, make sure you find out if they have been vaccinated and what type of vaccination(s) were given. This will guide you on whether you need to get any vaccinations done.

5. Find Out Any Specific Care Information for the Particular Breed You are Buying

Breeders can be a great source of information on caring for your birds, especially if the breeder has been doing it for years. Get information on things such as climate needs/ preferences, breed temperament, ideal feed, average egg production, and any other details you are unsure about.

The more you know about the chickens you are buying, the better equipped you will be to take care of them. So, do not hesitate to get as much information as you can from the breeder.

What to Look for When Buying Baby Chicks

The last thing you want is to buy chicks that are unhealthy or poorly developed. One infected bird can easily spread diseases to the rest of your flock. Therefore, before you take any chicks home, make it a point to check for any signs of ill health.

So, how do you know if the chicks you are buying are healthy? Here are some signs to look out for when buying chicks.

I. Eyes should be clear and alert.

II. Check the abdomen for any sign of distension.

III. The chick should be steady on their feet.

IV. The chick should be active and peeping.

V. Check that the top of the beak is aligned with the bottom.

VI. If the chick appears too small or runty compared to others, they may be in poor health.

VII. Healthy chicks are fluffy.

VIII. The vent should be clear of feces or any redness as this could point to diarrhea.

IX. Check to see that the brooder is clean.

What to Look for When Buying Mature Birds

A healthy bird is crucial, especially if you are just starting out in raising backyard chickens. Bringing home birds that do not lay eggs, thrive, or even end up dying can be a disappointing way to start your venture. This means being on the lookout for any signs of ill-health can save you lots of trouble later on.

When buying mature chickens, there are physical signs that can indicate that the chicken is in poor health. These include:

● Any discharge from the eyes or nostrils. A healthy chicken has clear eyes.

● Droopy or swollen eyes. A healthy chicken has clear eyes that are lively and alert.

● A hunched appearance. A healthy chicken has an upright gait and will not be hunched over.

● Wounds on the legs. The skin on the legs of the chicken should be free of wounds.

● Bald spots without feathers usually indicate that the chicken may have mites or lice.

● Crooked beak.

● Coughing or wheezing are signs that the chicken is sick.

● A droopy head is a sign of illness.

What to look out for when buying chickens for laying eggs:

If you are specifically looking to buy good egg-layers, then there are a few pointers that will help you identify chickens that have already started laying eggs.

● Check for bright combs and bright eyes. If a pullet has a dull comb, they have probably not reached egg-laying age.

● Chickens that are have already started laying eggs have wide hip bones as opposed to the narrow hips found in pullets that have not yet reached the egg-laying stage.

How Much Should You Expect to Pay?

Chicken prices will generally vary from breeder to breeder. Some breeds also tend to cost more than others, so all this will influence how much you will pay for the chickens you want. However, here are some guidelines with average indicative costs.

I. Chicks tend to be cheaper than mature birds. Chicks for most breeds cost anywhere between a dollar and five dollars.

II. Pullets that range in age from 1 month to 4 months (4 -16 weeks) will, on average, cost between $15 and $ 25.

III. Mature chickens or laying hens will range anywhere from $10 to $100 depending on the breed.

Baby Chicks or Hens – Which are Better?

When starting your backyard flock, you may be wondering whether to go for mature chickens or start with chicks. The choice will really come down to whether you are willing to wait the six months it takes for the chicks to mature and start laying eggs. Some people choose to go for mature chickens because they do not require as much care as baby chicks do. Ultimately, make a choice based on your circumstances and how well-equipped you are to take care of baby chicks.

Baby Chicks Pros
- Cheaper than mature chickens
- Require less feed
- Easier bonding with your pet

Cons
- Six months wait for eggs
- Require more attention and care

Buying pullets, which are adolescent hens of 15-22 weeks of age, may ultimately be better for you if your sole purpose for raising chicken is eggs. This is because pullets are usually just about to start laying eggs and will give you more than older hens.

If you do go for chicks, then you will need to have a brooder ready for them. A brooder is basically the first place where your chicks will go when they get home. It helps to keep them warm and well-insulated at their tender age. You do not have to buy a brooder; you can improvise using a container or a cardboard box. Just make sure that your brooder, whether store-bought or improvised, has at least 2 square feet of space for every chick.

It is important to make sure that your brooder is deep enough – at least 12 inches deep. This keeps your chicks safe and ensures they do not jump over the sides. You do not need to cover the brooder if it is deep enough. However, if you choose to cover it, make sure you use a breathable material since your chicks will need ventilation.

To maintain the temperature in the brooder at the required level, you will need a brooder lamp. You can buy a 250-watt heat lamp from most hardware stores or feed stores. The lamp should be mounted with a clamp to avoid fire hazards. Finally, you will need to put bedding on the brooder floor to keep the brooder moisture free and well insulated. Pine shavings are recommended as an ideal bedding material for brooders.

Once you have set up the brooder, you can now put in your chick feeder and waterer. Buy a feeder that is specifically designed for chicks since it will be easier for your chicks to use. It is best to have your brooder set up before the chicks arrive. Chicks tend to be delicate, and making sure that their brooder is well set up and ready for them will help you get off on the right foot.

Your chicks will stay in the brooder for about five weeks. After this period, they can be safely moved to the main chicken coop. You may want to keep them inside the coop for the first couple of days so that they can understand that the coop is their "home." Once they get used to the coop, they can be allowed some time outdoors in the chicken run.

Chapter 7: How to Feed and Water Your Flock

A healthy flock is a happy flock, and this will only be achieved if your chickens are fed on a wholesome and balanced diet. Chickens, for the most part, are not fussy eaters and will happily eat scraps from your table, bugs, and weeds on the ground and, of course, chicken feed. This means that feeding your chickens is not going to be too much of a hassle, provided you know what a healthy chicken diet consists of.

Feeding Chickens at Different Life Stages

The nutritional needs of a chick will vary from those of a mature hen or even a pullet. This means you need to feed your chicken's age-appropriate feed that will best meet the nutritional requirements for the stage of life they are in.

Starter Feed

Baby chicks, ranging from day-old chicks to 18 weeks old, require a starter diet. At this tender age, chicks need plenty of protein to promote growth and development. This is why starter feed is recommended for chicks as it contains more protein than any other type of chicken feed. On average, starter feed for baby chicks will have a protein content of 22%. This protein is essential for healthy growth and for the formation of feathers, which are predominantly constituted of protein.

Another reason you need to make sure that you are only feeding your chicks starter feed is that it contains low levels of calcium. Chicks are super sensitive to calcium, and if they consume high amounts of this mineral, it can lead to deformities in the bones and even cause kidney damage. Feed for layers, or mature hens, tends to have high levels of calcium, which is required for egg formation. While this is beneficial to the egg-laying hens, for baby chicks too much calcium is detrimental, and you should therefore never feed your chicks layer feed even for a day.

Your small chicks will have tiny beaks, so they require feed that is ground into fine pieces to make it easier for them to eat and digest. Starter feed is designed to be fine enough for baby chicks, so while your chicks are below 18 weeks of age, the best feed for them is starter feed.

When shopping for starter feed, you will notice that starter feed is available in both medicated and non-medicated options. If your chicks have been vaccinated against coccidiosis, do not feed them medicated starter feed. The medicated chicken feed contains amprolium, a compound that tends to affect the efficacy of the vaccine. On the other hand, if your chicks have not been vaccinated against coccidiosis, the amprolium in medicated feed serves to protect your chicks from the disease.

Keeping your chicks in hygienic conditions helps to boost their natural immunity, so you do not necessarily need to feed your chicks on medicated starter feed to keep them safe from diseases.

Grower Feed

From eight weeks old, your chicks need grower feed, which is designed to keep your chicks growing until they reach laying age at between 18 t0 22 weeks. Since chicks in the growing stage, between 8 to 18 weeks, are not yet laying eggs, grower feed contains less calcium than layer feed. The protein content in grower feed is not quite as high as that in starter feed, but it is sufficient to help your chicks mature into layers.

Just like baby chicks, pullets need age-appropriate food because it meets their nutritional needs best. Do not feed your pullets or growing chicks layer feed because it contains too much calcium for that age and may cause health issues in the long run.

Layer Feed

Layer feed is appropriate for chickens over 18 weeks old that have started laying eggs. Layer feed is designed to provide all the essential nutrients needed to keep your mature chickens healthy and productive in terms of egg-laying. Layer feed contains more calcium than either the grower or starter feeds, and this is because laying hens need more calcium for proper shell formation.

The protein content in the layer field is at about 16%, and while this is sufficient to meet the nutritional needs of a mature chicken, it would be too little to meet the needs of baby chicks or growing pullets. That is why it is important to only feed layer feed to adult hens of 18 plus weeks.

Broiler Feed

If you are raising your chickens for meat purposes, the recommended feed for them is broiler feed. This type of feed is rich in protein and is formulated to promote faster growth and boost weight gain. It helps your chicken to put on weight fast, which is a desirable characteristic in birds bred for meat purposes.

Do not feed broiler feed to layers since it does not have the necessary nutrient content to boost egg production.

Different Forms of Chicken Feed

Chicken feeds are available in different forms. They can be in the form of mash, pellets, or crumble. Mash is a fine loose chicken feed that is easy to digest. You will find that chick feed comes in the form of mash since this is the easiest form for small chicks to digest. Grower feed also often comes in mash form as does layer feed.

Crumble is a semi-loose type of chicken food. It is coarser than mash and can be fed to either pullets or layers. Apart from crumble, chicken feed is also available in the form of pellets. Pellets are popular since they tend to be less messy compared to mash, and most people find them to be easier to handle.

Supplements

• Crushed Oyster Shells

Laying hens may need an additional source of calcium in addition to what is in their feed. That is why crushed oyster shells are recommended for laying birds. The supplemental amount of calcium helps to boost egg production and shell formation. The oyster shells do not need to be mixed with the chicken feed, simply provide them in a separate trough or feeder.

Chickens can control their calcium intake based on what they need, so you do not need to worry about providing them with too much shell grit. They will only eat what their body needs.

• Grit

Grit is used to refer to hard materials such as sand, small stones, or dirt that are provided to chicken to aid in digestion. Chickens need grit in their diet to enable them to digest fibrous foods such as grains in their gizzard. If your chickens spend time outdoors, they will pick up grit from the ground as they scratch and dig around so they will not need additional grit in their diet. However, if your chickens are confined to their coop, you need to provide them with some grit in a separate container to enable them to digest fibrous foods.

Chicks and pullets that are only fed on starter feed or growers feed do not require any grit since they are not fed grains or other foods that are hard to digest.

• Treats and Scraps

Human food is generally safe for chickens, and there is no problem feeding chickens scraps from your table. However, since they are getting all their nutritional requirements met by their chicken feed, it is recommended that you keep treats and table scraps to a minimum. Avoid feeding your chickens fatty foods as this may lead to obesity and, in some cases, even hinder egg production.

Treats such as chicken scratch can be used to boost your chickens' carbohydrate intake. Chicken scratch typically contains a mixture of different grains. Though grains are good for your chicken, they do not contain all the nutrients that your birds need, so always use chicken scratch as a treat and not the main staple of your chickens' diet.

Too much chicken scratch can lead to obesity since it is high in carbohydrates. Chicken scratch also lacks all the nutrients your chickens need, so relying on it as your primary chicken feed is not recommended. As long as you are feeding your chickens the appropriate feed, treats and scraps may be given occasionally, but they are not necessary.

Free Feeding vs. Restricted Feeding

Chickens eat pretty much all day, so restricted feeding is not really recommended. A chicken can only eat small bits of feed at a time, so in most cases, they will eat a little at a time throughout the day. By using feeders that replenish feed as it is eaten, you can ensure that your chickens have access to feed through the day without necessarily having to keep replenishing their feed manually.

Finding the right feeder for your chickens will make the feeding process so much easier. Feeding chickens on the ground is not recommended, as when this is done, the feed ends up mixing with chicken poop and other kinds of dirt in the ground. This can lead to diseases and infections in your flock. To avoid this, an appropriate feeder will come in handy.

Automatic Feeders

Automatic feeders are convenient, easy to use, and also help to reduce the waste of feed. With this kind of feeder, you will store your chicken feed in it, eliminating the need to keep refilling your feeder.

With an automatic feeder, the feed is dispensed as needed, so you may end up saving on chicken feed costs. The downside to this is that since chickens can access the feed anytime, it can encourage overeating, so there are both pros and cons to using automatic feeders. These types of feeders are also effective in keeping pests and bugs away from the chicken feed. However, an automatic feeder is typically more expensive than other types of feeders.

Ultimately, if you do not fancy having to feed your chickens manually every other day, an automatic feeder is the way to go.

Gravity Feeders

These feeders are simple to use and operate by simply releasing feed downward as it is eaten. You can mount a gravity feeder or leave it as free-standing, depending on where you choose to position it. Unlike the automatic feeder, this type needs to be replenished often since you can only put in a limited amount of feed.

The number of feeders that you will need will depend on the size of your flock. Aim to have at least one feeder for every ten birds. If you have birds of mixed ages in the same coop, you should have a separate feeder for your chicks to ensure that they are only eating their feed and not layer feed.

Gravity feeders are inexpensive, easy to use, and a convenient option if you have a small flock.

Watering your Chickens

Mature chickens drink approximately a pint of water daily. Since they drink this amount in small portions throughout the day, it is essential to make sure that your chickens have full-time access to clean water. Lack of enough drinking water can cause poor egg production, ill health, and even poor development.

Waterers come in handy because they help to deliver water efficiently to your chickens. When you provide water to your chickens in open containers, chances of dirt and debris contaminating the water is high. This is why a waterer is more suitable and hygienic for your flock.

Galvanized Waterers

When using a galvanized waterer, vacuum pressure allows water to keep filling the drinking trough as needed. This limits waste and prevents overfilling. However, you will need to put the waterer on a level surface for it to function properly. Alternatively, you can also suspend or hang your waterer from the roof of the coop.

A galvanized waterer is typically made out of steel and is therefore very durable. However, if you plan to supplement the water with vinegar or other supplements, they will react with the metal, so it is best if you go for a plastic waterer.

Plastic Waterers

Just like galvanized waterers, plastic waterers release water as it is needed. This helps to eliminate waste and also keep the drinking water clean. Plastic waterers come in a variety of sizes ranging from small chick-appropriate sizes to larger ones. This type of waterer is easy to use and is the most popular type of waterer among people who raise backyard chickens.

With this kind of waterer, you can add supplements to the water since the supplements will not react with the plastic. This type of waterer is also great for extreme heat conditions since it will not heat up as fast as galvanized metal. Plastic waterers also insulate the water better than metal waterers in cold temperatures.

Nipple Waterers

Nipple waterers typically have little plastic nipples or outlets attached to the main waterer so that instead of drinking from a trough or lip, your chickens drink from the nipple. These waterers help to keep the mess to a minimum. However, you will need to train your chickens to drink from this type of waterer until they get the hang of it.

Some waterers will have cups instead of nipple outlets. These can be bought separately to attach to your normal waterer, or you can buy a waterer that already has them attached.

Homemade Waterers

You can easily fashion a homemade waterer for your chickens using a plastic bucket and a dish. Simply drill some holes on the bucket. Drill the holes lower than the top of the plastic dish you will be using. Fill the bucket with water and replace the lid. Your bucket should sit on top of the plate allowing a drinking area for the chickens along the edges.

Watering in Winter

When temperatures drop, water tends to freeze, so you will need to make sure that your chickens still have access to water during winter. Replenishing your waterers with warm water often is one way to ensure that your chickens have access to drinking water during the cold season.

If you are using a galvanized waterer, having a heat lamp directly over it can help to prevent the water from freezing. Some waterers come with heated bases that can be connected to electricity to keep the water from freezing. This will also help in ensuring that your chickens have access to water during the cold winter months.

Signs of Poor Nutrition in Chicken

For your chickens to remain healthy and productive, a healthy diet is crucial. Taking care to observe any signs of nutritional deficiencies in your flock will guide you in knowing whether you are feeding your chickens properly.

Here are some classic symptoms of poor nutrition that you need to be on the lookout for.

I. A drop in egg production

II. Poor feathering

III. Eggs with thin shells

IV. Curved legs

V. Stunted growth

VI. Ruffled feathers

VII. Toes that curl inwards

VIII. Chickens eating their own eggs

Proteins, vitamins, minerals, and carbohydrates all play a crucial role in ensuring good health in chickens. Always ensure that your chickens are getting the right feed that is age-appropriate. If your chickens are enclosed or restricted to an area where they cannot forage on the ground, you can include supplements in their diet to make up for any nutrient deficiencies in their food.

Ultimately you will only get the best out of your chicken if it is well fed and cared for. Healthy birds produce more eggs than those that are in poor health. Always check labels of the feeds you are buying to ensure that you are getting feed that meets the basic nutritional requirements of your chickens.

Chapter 8: Managing Your Laying Hens

With the increase in awareness of the importance of healthy food from healthy food sources, it is no surprise that an increasing number of people have turned to backyard chicken farming. Whether you keep a modest flock or dozens of birds, one thing most people agree on is that having your own supply of fresh eggs is pretty convenient. However, to get the best out of your layers, you need to ensure that they are well taken care of.

Start with Healthy Chicks

If you are raising backyard chickens for the eggs, you can either buy mature hens or chicks. Chicks tend to be cheaper to buy compared to adult hens. However, there will be a waiting period before you can start collecting eggs. Chicks may mean a lot more work in terms of care and maintenance, but once they start laying eggs, they are likely to produce more eggs than adult hens. On the other hand, chicks do require a lot of care, so if you do not have a lot of time for care and maintenance, you can always buy adult hens.

If you choose to start your flock with chicks, the kind of care they get at the early stages of life will definitely impact their egg production in adulthood. One common mistake you need to avoid is feeding baby chicks food for layers. Even if your chicks are meant to be bred into egg-layers, they should never be fed layer feed until they are at least 18 weeks old.

Layer feed has high levels of calcium that is beneficial for egg-laying hens since it helps in shell formation. However, chicks do not require high calcium levels, and if they consume too much of it, it can lead to kidney problems and bone deformities. Always feed your chicks on starter feed until they are 18 weeks old. After 18 weeks, most breeds will be ready to start laying eggs, and at this point, you can safely switch them from starter feed to layer feed.

Your chicks should always have access to clean water. You can have your chick waterer suspended above the floor of the brooder so that it does not get contaminated by poop or any other dirt. You must always keep the brooder clean if you want them to stay healthy. If you leave the bedding in your brooder too long, the accumulation of poop and moisture can lead to diseases.

A dirty brooder can undermine all your hard work even if you are feeding your chicks the right feed. Clean your brooder as often as possible, and do not let the bedding get moist. A dirty brooder can lead to diseases and negatively impact the proper growth and development of your chicks.

Once the chicks have started to grow some feathers, usually in about five-to-six weeks, they are ready to go to the main coop. From this age on, they can be allowed outdoors, although you will need to make sure that they are safe from predators and rodents.

Feeding your Layers

When your chickens reach the egg-laying stage, their nutritional needs evolve to enable them to produce eggs. This means that they need to be fed layer feed. This feed has a healthy dose of calcium, which is required for shell formation. It is important to ensure that your layers are eating the right food and that they have enough of it.

Chickens tend to eat throughout the day. The best way to feed them is through feeders that dispense food as it is eaten. This helps to ensure that your layers have access to food whenever they need it. The primary source of nutrition for your layers should be layer feed. While chickens are happy to eat anything, including scraps from your table, they do require a nutritional diet, which is only achieved by feeding them mostly on high-quality layer feed.

Even with high-quality layer feed, your layers will still require an additional source of calcium. That is why it is important to provide your chickens with crushed oyster shells. Crushed oyster shells are a great source of calcium for layers. All you need to do is put them in a separate dish or container when feeding your chickens. You do not need to worry about your chickens eating too much of the oyster shells. Chickens will eat just as much calcium as their body requires. Make oyster shells a staple in your chicken's diet if you want to boost proper egg production.

If your chickens are not free-range, it means they may not be getting enough grit. Grit is coarse material that chickens ingest from the ground to aid in the digestion of fibrous materials such as grains. For enclosed birds, you will need to provide them with grit to complement their diet and aid in proper digestion. Do not mix the grit with the regular chicken feed but rather provide it separately. Just like the oyster shells, chickens will only ingest as much grit as they require, so you do not need to worry that they may eat too much of it.

Apart from their main layer feed, here are some treats that you can include in your chickens' diet to keep them laying eggs.

- **Mealworms**

This treat is full of healthy protein and makes a very healthy treat for chickens. It also contains plenty of essential minerals and vitamins that are good for your chickens' health. However, do not feed too much of it to your layers as they do not require excessive amounts of protein. A spoonful of mealworms per chicken once or twice a week should be sufficient.

- **Cracked Corn**

This is a healthy treat for layers. Corn is, however, high in carbohydrates, so it should be fed to chickens in moderation to avoid obesity. Excess weight gain reduces egg production and is not good for your chickens' health.

- **Greens**

Leafy greens and vegetables have plenty of essential minerals and vitamins that help to keep your chickens healthy. Kale, cabbage, and dandelions all make great healthy treats that you can feed to your chickens occasionally.

Fruits like watermelon are also good for your chickens and can be given as treats every so often.

- **Scratch Grains**

You can feed your chickens scratch grains as treats provided you do so in moderation.

- **Scraps and Leftovers**

Chickens can safely consume human food, as most human food is also safe for chickens. However, when giving your chickens scraps from your table, avoid certain foods, including avocado, tomato stems, and fruits like lemon and oranges. Foods like garlic and onions should also be avoided. Bear in mind that table scraps should be given in moderation as they can lead to obesity, which in turn will affect the health of your chickens.

Finally, your layers need to have access to clean water at all times. Find a suitable waterer to use and ensure that you always keep it clean. Contaminants from chicken poop, debt, and debris can easily contaminate water. If you find that the chicken water has dirt in it, pour it out and replace it with clean drinking water.

Housing

Your chicken coop and chicken run need to be kept clean to ensure the health of your chickens. Make sure you have suitable bedding to keep the coop moisture-free. Bedding also helps to prevent the accumulation of ammonia in the coop. If ammonia from chicken manure builds up to very high levels, it can lead to respiratory diseases, so it is best to ensure that your chicken coop is well ventilated.

Layers need a private space for laying eggs and for brooding. You should have nest boxes in your coop where your layers can lay eggs. The nest boxes need to be cushioned with bedding. Straw and hay make great bedding for nest boxes since they are soft. This bedding will cushion the egg once they are laid and also help to insulate the chicken. However, just like the bedding in the rest of the coop, change the bedding in the nest boxes often to keep them clean. Nest boxes should be cleaned at least once a month.

Nest boxes need to be slightly raised off the ground. Your nest boxes should be dimly lit, so just ensure that they are not positioned in an area with direct sunlight. Some people use curtains, but this is not necessary provided you have positioned the nest boxes in a quiet area of the coop.

You need at least one nest box per four chickens so that your layers all have access to a next box when they need to lay eggs. If the nest boxes are too few, your chickens may resort to laying eggs in hidden nooks and crannies that may be hard for you to reach. Most of all, make sure that your nest boxes are safe from predators, rodents, and other pests. Eggs can attract predators, so your nest boxes need to be checked regularly for pests and rodents such as mice.

Winter months pose a challenge for chickens since the cold weather can affect egg production if your flock is not comfortable and properly insulated. To make sure that your layers are comfortable during the cold season, here are the factors you need to keep in mind.

1) Light

Layers need light to stimulate the pineal gland. This gland initiates egg production by releasing hormones to initiate the process. This means that your chickens need daylight to produce eggs. In winter months, you can substitute daylight with a 60-watt incandescent light bulb. Make sure you provide light for at least 16 hours each day.

2) Roost

Roosting bars are an essential part of any coop. Your layers need a comfortable area to perch on, and roosts provide them with this space. When it is cold, chickens tend to roost close together to keep each other warm. Ensure that coop has sufficient roosting space for your layers. A general rule of thumb is to have at least 8 inches of roost space per chicken.

3) Keep the Water Supply from Freezing

Water tends to freeze in winter, especially if it is in galvanized waterers. This means you will need to keep a fresh supply of warm water to your layers during colder months. Chickens will not lay eggs if they do not have access to sufficient water, so making sure that their water is not frozen over is crucial.

4) Deep Litter Method

Bedding does not just help to keep the coop clean and odor-free; it also helps to insulate the coop, making it warmer and more comfortable for your chickens. During winter having a deeper layer of bedding and litter than you do during summer months can help to keep the coop warm.

To use the deep litter method to keep the chicken coop warm enough in winter, you can simply keep adding on to your regular bedding as winter approaches by adding layer after layer of bedding periodically. By winter, if your litter is up to 8 inches deep, the lower layers of bedding will start giving off heat as they compost, and the coop will be much warmer.

5) Provide Warming Treats

Treats such as corn, which boost metabolism in chickens, can help to keep the chickens warm. You can feed cracked corn to your chickens in the evenings to keep them warm during the night.

Reduce Stress for Better Egg Production

Winds, extreme heat, and cold snaps are all stress factors that can affect your chickens' ability to produce eggs. To reduce the levels of stress that your chickens go through during harsh conditions, here are some easy tips to help keep egg production up and your chickens healthy.

• Boosting protein intake can help to minimize the effects of stress on your chickens. You can add protein-rich treats such as mealworms to your chickens' diet.

• Green feeds such as vegetables can help to boost fertility and egg production in chickens. They are rich in essential vitamins and minerals and will have beneficial effects, especially during high-stress seasons.

• Adding vitamins or supplements to drinking water may also help in boosting egg production.

• Heat stress is also bad for chickens and can cause a decrease in egg production. Ensure that during super-hot months your chickens have access to shaded areas where they can cool off.

Reasons Why Chickens Stop Laying Eggs

1. A poor diet is one of the primary reasons why chickens stop laying eggs. Always feed your layers the recommended feed for egg-laying chickens and include healthy treats such as oyster shells to boost calcium levels in the diet.

2. If your chickens are not getting enough daylight, they may stop laying eggs. Chickens need at least 16 hours of daylight to produce eggs. An artificial light source can help you ensure that your chickens get sufficient hours of light every day.

3. Brooding hens do not lay eggs. They will typically spend a lot of time in the nest box and may become protective over their space since they are trying to hatch eggs. This process usually lasts for 21 days.

4. Some chicken breeds are not as prolific layers as others, and may only lay two or three eggs in a week.

5. Diseases and parasitic infections can interfere with egg production, so if your chickens are in ill-health, the chances are that their egg production will drop.

6. Chickens will eventually stop laying eggs due to old age. Most chickens will actively lay eggs for about three years, but after that, there will be a natural decrease in egg production until it ceases altogether.

Chapter 9: Understanding Chickens

As a first-time owner, you may observe behavior in your chickens that you may not understand. Chickens tend to vary in terms of temperament, preferences, quirks, and even activity levels. Your chickens will have different personalities, and so it is not unusual to find that your flock is composed of chickens that each have their unique characteristics. Understanding why your chickens behave a certain way may help you in caring for them and bonding with them.

Chickens do better when raised in groups or flocks. This is because chickens are naturally social, so they thrive in groups or flocks where they are part of a family or community. Watching your chickens as they interact and go about their business can be quite interesting, and many people who enjoy raising chickens as a hobby often find themselves enjoying "chicken TV". However, it is important to know what behaviors constitute normal chicken behavior and which may be signs of illness or stress.

Normal Chicken Behavior

Pecking Order

In every flock, there is a pecking order. This is just the way the social hierarchy of chickens works. Whenever chickens are put together in groups, they will find ways to establish ranks where there is a kind of social structure, and everybody knows their place. This happens even amongst baby chicks, and you will find that even birds at that tender age have a pecking order.

Chickens will often fight each other to establish and maintain a pecking order, so do not be surprised to see the chickens in your flock engage in squabbles from time to time. Most of the fights are usually short-lived and do not really end up causing grievous harm. This, however, may not be the case if your flock has several roosters. Roosters can fight to the death, especially if there are hens in the flock to fight over.

Small flocks will tend to have fewer squabbles over rank for the simple reason that it is easier to establish a pecking order in smaller groups. Larger flocks will have more frequent squabbles, especially if the flock has several roosters. In any flock where there is only one rooster, he will dominate the hens and will be the default leader of the flock.

The lead rooster maintains the social hierarchy in his flock and even comes to the rescue when the hens in his flock are squabbling. Although the rooster becomes protective of all of the hens in his flock, often, he will have a favorite hen whom he obviously favors over the others. This kind of favoritism is a mating ritual of sorts, and the rooster will mate more often with his favorite hen than with the other hens in his flock.

In flocks without a rooster, a dominant hen becomes the leader of the flock. She will be in charge of the flock and will take on the role of protector and peacemaker for the rest of the flock. The pecking order is usually disrupted when new chickens are brought into the flock. If you introduce new birds to an existing flock, there is likely to be a certain amount of squabbling. This is usually a way to show the new birds who's in charge and establish their places or rankings in the flock.

To keep squabbles to a minimum when you bring new birds home, you can separate them from the rest of the flock for a day or two. This will give them time to get familiar with each other without necessarily being within fighting distance.

Ultimately, squabbles and fights between chickens are normal. It is simply how they establish their pecking order. You do not need to try and separate them during such fights since these squabbles are usually short, and in most cases, no blood is shed. However, roosters can kill each other in the course of serious fighting, so once they start to draw blood, you should separate them to avoid a fatal outcome.

Dust Bathing

You will often notice that your chickens love bathing themselves in the dust. Dust bathing is a normal behavior in chickens, and providing a dust bath in your chicken run is actually recommended in order to keep your chickens happy. When taking a dust bath, your chicken will find a spot with loose soil. They will then dig a depression in this patch before sitting in it and throwing the soil over themselves with their feathers and legs.

Dust bathing helps chickens to get rid of mites, lice, and other parasites. It is also a fun experience for them, so if your chickens are confined to a run, you can always provide them with a box of sand for dust bathing.

Broodiness

Hens will get broody every so often. This is when the hen wants to hatch eggs. Brooding hens become inactive and may sit in the nest box for days on end. They may also become aggressive or protective as they try to guard their nesting space. If you want chicks, this is the best time to put eggs in your chicken's nest box and let her hatch them. However, if you do not want chicks, you can stop a chicken from being broody by lowering their body temperature. A cold bath or keeping them from the nest at night are two easy ways to do this.

Crowing

Roosters crow every day. This is just part of their nature; they will crow at the crack of dawn and throughout the day. This is part of the reason that most cities and towns ban the keeping of roosters. The crowing is, unfortunately, not something you can stop, since it is a natural part of a rooster's behavior.

When crowing, roosters are essentially making their presence known to the other roosters, to the hens around them, or simply just expressing themselves. Even if your area allows you to raise roosters in your backyard, be sure that you are ready to contend with them crowing because it is going to be an everyday occurrence.

Preening

You may notice that your chickens spend quite a bit of time pecking at their feathers. This kind of preening is part of normal chicken behavior. When preening, chickens are essentially removing dirt, pests, or bugs from their feathers, so in effect, preening is your chicken's way of keeping themselves clean.

Molting

Chickens usually undergo a period where they shed old feathers and grow new ones. This process is referred to as molting and will typically occur when temperatures start getting cooler. During this molting period, you will notice that your chickens will stop laying eggs as they are reserving their nutrients for the feather renewal process.

The molting period tends to vary from chicken to chicken, but on average, it will range from 4 to 16 weeks. You can help to speed the process up for your chickens by boosting their protein intake. Feathers are mostly made up of protein, so the higher the amount of protein in your chicken's diet, the faster the molting process will be. It is also best to avoid stressing your chickens during this period. This means you should not handle or touch them as their bodies are super-sensitive during the molting period.

Scratching and Digging

Chickens love to scratch and dig the ground. They dig up bugs, worms, and grit to eat from the ground. You will notice that your chickens will spend most of their time outdoors digging and scratching. This is normal and expected chicken behavior, and it is recommended that you provide your chickens with a foraging area outdoors where they can dig and scratch. If your chickens are not free-range, you can confine them to a run, which will give them foraging space where they can scratch and dig.

Celibacy

Hens do not need a rooster in the flock to be happy or to lay eggs. Hens that are raised without roosters create their own society and pecking order where the dominant hen becomes the head of the group. Hens will lay eggs as they normally would without a rooster in the flock. However, since these eggs will not be fertilized, they will not be able to hatch chicks from them.

Abnormal Chicken Behavior

There are chicken behaviors that should serve as an indicator that there is a problem with your flock. Abnormal chicken behavior can be caused by illness or stress factors, so observing your chickens often will help you pick up on any unwanted behavior. Here are some abnormal chicken behaviors for which you need to be on the lookout.

Aggression

Squabbles and fighting, as we have learned, are acceptable and normal behaviors that chickens use to establish a pecking order in the flock. However, in some cases, you may have overly aggressive birds that attack you or your children. This behavior is commonly observed in roosters. Roosters may get into a habit of charging anyone who comes near their space. They will peck, slap with their feathers, and try to hit with their claws or spurs.

This kind of aggression can be dangerous, especially if you have children, and needs to be addressed. A rooster that attacks humans is trying to establish dominance over them, and if the behavior is not curbed, it can become a serious problem. To establish that you are boss, you need to handle the rooster a bit forcefully. This means that if he pecks at your feet, you should shove him with your feet and force him directly to the ground.

The aim here is not to hurt the bird but rather to force it into a submissive position. You can also hold him down for a few minutes. Forcing the aggressive rooster to stay still is a way of establishing who is in charge. Roosters will generally stop attacking humans once they are made to understand that humans rank higher in the pecking order.

Hens will rarely be aggressive toward humans unless they are protecting their baby chicks. This is just a natural protective instinct and will only last while the chicks are young. Avoid touching or handling baby chicks as this may cause the mother hen to feel that her babies are in danger and attack.

Pecking and Feather Picking

When your chickens do not have enough space and are in an overcrowded coop or run, they may resort to incessant pecking and feather picking. This can be a sign of stress or boredom. Always ensure that your run and coop have enough space for the number of birds you have in your flock. The recommended space in the coop per chicken is at least three square feet while your chicken run should have at least eight to ten square feet of space for every bird.

Wing Dropping

If you observe your chickens dragging their feathers on the ground, this is a common sign of illness. Droopy wings can point to any number of conditions, and you may need to have your chicken checked for diseases. Remember that chicken diseases can spread pretty fast among the flock, so early intervention can help you prevent a disease from spreading.

Lethargy

A normal chicken is by nature alert, active, and curious. They will be scratching and digging at the ground, moving around constantly, and interacting in one way or another with the rest of the flock. If your chicken appears dull or inactive and exhibits general disinterest in what is going on around them, they may be unwell. If you observe that the chicken is having trouble holding its head up or walking, this is a clear sign of ill-health, and you need to have a vet look at it.

Hens Eating Their Own Eggs

When hens start eating their own eggs, this can become a serious problem. This behavior is commonly caused by a calcium deficiency, and the chickens start to eat eggs as a way to supplement their calcium intake. If you let this behavior go on for a long time, it will become even harder to break, so it is best to stop it as soon as you realize that it is happening.

To stop this habit, feed your chickens extra calcium by providing them with crushed oyster shells. Avoid feeding them their eggshells unless they are completely crushed; otherwise, they will start to associate their eggs with the shells you feed them.

Another reason that may encourage hens to eat their eggs is egg breakage. Once an egg breaks, then the hens will be likely to eat it. To avoid this, make sure that your nesting boxes are well cushioned with straw and hay. You also need to avoid congestion in the nesting boxes by having at least one nest box for every four chickens in your flock.

Ultimately chickens have unique quirks and personalities, and the best way to understand your flock is to spend some time watching them. In this way, you will get to know what the normal behavior for them is. Once you understand their routine behavior, it will become easier to identify any uncharacteristic behavior that may be caused by stress, illness, or other factors. A good chicken farmer knows their flock well and makes it their business to understand their chickens.

Chapter 10: All About Eggs

One of the main benefits of raising chickens in your backyard is the fresh supply of eggs. As more and more people look for ways to produce their own food and take control of what kind of food is on their table, the popularity of raising chickens keeps increasing. For beginners who have just started raising chickens, collecting your first batch of eggs can be quite satisfying.

Bright yolks, firm whites, and of course, tasty goodness are some of the hallmarks of fresh eggs. When you compare eggs from your coop with grocery-bought eggs, the difference is usually quite clear. When you purchase eggs from a grocery store, you have no way of knowing just how fresh they are, how the chickens that produced the eggs were raised, or what kind of feed they were given.

When you are raising chickens in your backyard, you have control over their diet. This means you can choose to go organic and, in this way, ensure that your eggs are completely natural and GMO-free. This is what makes raising backyard chickens so fulfilling. If you are a beginner, you will soon realize that eggs come in different shapes, colors, and even sizes. From brown eggs to white and even blue eggs, there is an array of not just colors but qualities of eggs.

Before you can get to enjoy your eggs, you first need to know the best practice when it comes to collecting cleaning and storing them.

Collecting Eggs

When your hens lay eggs, you do not want to leave them lying around in the coop for too long. Here are some of the reasons why collecting your eggs regularly is important,

- Eggs are fragile and the longer you leave them in the coop, the higher the chances of them getting trampled and broken.
- Eggs can attract predators and rodents to the coop. Cats, raccoons, rats, and other types of predators like the taste of eggs so they can get into the habit of getting into the chicken coop if eggs are constantly left lying about.
- Eggs do not have a very long shelf life, so if you want to enjoy your eggs fresh, it is best to collect them often from the coop.
- Hens can start to eat their own eggs if they are not collected often. This happens especially when there are broken eggs in the coop, and hens get into the habit of eating them.
- Coops tend to have plenty of contaminants in the form of chicken manure. You do not want your eggs to stay too long in the coop. The longer the eggs stay in the chicken coop, the more likely they are to get contaminated with dirt and chicken poop.

If you have a small flock, collecting eggs once in the morning and later in the evening is advisable. People with large flocks should collect eggs thrice a day. This will ensure that eggs laid during the day do not stay in the coop overnight. A plastic container should be sufficient when collecting eggs. Just be sure not to stack them too high to avoid accidental breakage.

Cleaning Eggs

When chickens lay eggs, they usually have a natural protective layer on them to keep them germ-free. However, it is normal for eggs to get some dirt on them in the coop, so cleaning them before storage is good practice. In most cases, it is recommended that you clean your eggs with a dry piece of cloth. Using a dry cloth will help to get the eggs clean without damaging their natural protective outer layer.

Alternatively, there are times when your eggs may have poop stains and other kinds of dirt that needs to be washed out. In such cases, it is okay to wet clean the eggs with some water. Ideally, when using wet cleaning, you should use warm water. Once the egg is clean, you can dry it with a paper towel and then place it on a rack.

Always ensure that the nest boxes and coop are kept clean, as this will reduce the chances of collecting dirty eggs. Clean the bedding in the nest boxes as often as possible, and this will give your hens a clean place to lay their eggs. Ultimately this means cleaner eggs for you.

Storing Your Eggs

Whether your eggs are simply for domestic consumption or sale, proper storage is important in preserving freshness. Once your eggs are clean, they should be stored in an egg carton. It is recommended that you indicate the date of collection for the eggs on the carton so that you know which eggs are the freshest. This is especially important if you collect a lot of eggs from your flock daily. If you do not separate them by date, you risk some of them going stale on you.

Always use eggs in the order in which they were collected. This prevents situations where some eggs go bad because they have been stored for too long. As a general rule, store your eggs in the refrigerator. Refrigerated eggs will, on average, have a shelf life of one month from the date of collection. Eggs that were not wet-cleaned after collection can last for several weeks stored at room temperature. Always wash your eggs before using them to get rid of any dirt or contaminants on the surface.

If you have stored your eggs for a while and are not sure whether they are still fresh, you can use a simple float test to find out. Fill a bowl with clean water, then place the egg inside the bowl; a fresh egg will sink to the bottom, while a stale egg will float on the water.

Determining Egg Quality

The quality of an egg is typically based on internal egg quality and external egg quality. External egg quality centers upon the external characteristics of the eggs, such as cleanliness, shape, and even texture. If you plan to sell your eggs, they need to be graded as A or AA. If your eggs are graded as B, they are not approved for sale in stores.

External quality starts with how clean your eggs are. Even though a chicken will lay an egg when it is nice and clean, eggs easily get dirty in the nesting box. This is why it is important to collect your eggs as often as possible to keep contamination to a minimum. You can always dry clean or wet clean your eggs to keep them clean though this will affect their shelf life.

The other aspect that affects the external quality grading of an egg is its shape. Eggs that are any shape other than oval are considered to be lower quality. This does not mean that their nutritional content is any lower; it simply indicates that their physical shape differs from the ideal oval shape of an egg. Similarly, eggs with rough or uneven shells are downgraded since they are more likely to break than those with smoother shells.

Interior quality is graded based on the quality of internal features of the egg, such as the yolk. When an egg is fresh, the egg yolk tends to be round and firm. However, as time passes, the yolk starts to absorb water from the egg white and increases in size. This means that the longer an egg is stored, the more its internal quality reduces.

The internal quality of an egg is not just affected by the passage of time but will be affected by a host of other factors. These include disease, temperature, humidity, and storage of the egg. This means that to get high-quality eggs, your hens need to be healthy and fed on a well-balanced diet. How you handle the eggs and store them may also cause the internal quality of the egg to deteriorate.

When eggs are stored in high temperatures, the internal quality of the egg is reduced. This is why refrigeration is recommended in order to keep the eggs fresh for as long as possible. Rough handling may also interfere with internal egg quality, so always be gentle when collecting, cleaning, and storing your eggs.

Enjoying Your Eggs

Eggs are some of the most versatile foods on the planet. From breakfast to dinner and even desserts, eggs are a staple in many homes. They are used to create a wide variety of dishes. Starting with your morning omelet, your favorite pastry, salad, and many other meals, you are likely to find that eggs are ingredients in many staple dishes in a lot of homes. This is what makes having your own supply of fresh eggs so rewarding. Every time you use an egg from your backyard chickens, you can be sure of the freshness and quality of that egg.

What exactly is in an egg, and what makes this superfood so popular across the globe? Let's have a look at the nutrient content of an egg (boiled egg, values per 100 grams)

- Total fat 11 g (16%)
- Saturated fat 3.3 g (16%)
- Polyunsaturated fat 1.4 g
- Monounsaturated fat 4.1 g
- Cholesterol 373 mg (124%)
- Sodium 124 mg (5%)
- Potassium 126 mg (3%)
- Total Carbohydrate 1.1 g (0%)

- Dietary fiber 0 g (0%)
- Sugar 1.1 g
- Protein 13 g (26%)
- Vitamin A (10%)
- Vitamin C 0%
- Calcium 5%
- Iron 6%
- Vitamin D 21%
- Vitamin B-6 5%
- Cobalamin 18%
- Magnesium 2%

The humble egg carries a host of essential vitamins and nutrients, including protein. Eggs are also relatively low in calories, and this means they go well even with calorie-restrictive diets. Eggs are, in fact, a popular menu item for keto-dieters, so you should not fret too much about gaining excess weight from eating eggs.

If you have a constant supply of eggs from your backyard flock, remember there is plenty you can do with eggs in the kitchen. Try exciting new recipes, use them for baking your baked treats, and pretty much use your eggs as creatively as possible. There are plenty of egg recipes available online, so if you are looking for new ways to enjoy your eggs, there is always a recipe that you can try out and enjoy.

Chapter 11: Meat Birds

An increasing number of people are raising chickens for meat purposes. This is because as people become more sensitized to harmful chicken farming practices, they are choosing to have more control over the kind of food they eat. Factory farmed chickens are often not given the best care or feed, and raising your own birds for meat will give you access to more wholesome chicken meat.

You can easily raise chickens for meat in your backyard as the process of raising meat birds is pretty much the same as what you would do with any other chicken. The only difference usually comes in the type of feed that you will give your chickens if you are raising them solely for meat purposes.

The Best Meat Chicken Breeds

Meat birds vary from layers in that they tend to grow faster and also put on more weight. This means that while any chicken can be raised for meat purposes, ultimately meat breeds will give you more meat in a much faster time frame than layers. Here are some of the best meat chicken breeds that should be part of your flock if you are raising backyard chickens for meat purposes.

- **Jumbo Cornish Cross**

This is a large chicken breed that puts on weight pretty fast. They have large breasts and big thighs that have made them popular among chicken meat breeders. In about eight weeks you can expect a male Jumbo Cornish to weigh about four pounds, while a female will weigh in at two pounds at the same age.

- **Cornish Roaster**

This is another large chicken breed that is ideal for meat purposes. It has yellow skin, and, like the Jumbo Cornish, large breasts and thick thighs. This chicken breed matures fast, reaching maturity in about ten weeks.

- **Jersey Giant**

As the name implies, this is a large bird that is a favorite for many chicken meat farmers. It also has good egg production so it can serve as both a layer and a meat bird in your flock. Jersey Giants do not mature as fast as other meat birds, but it will grow to a considerable size and weight.

- **Freedom Rangers**

Freedom Rangers is another breed that is perfect for those who want to raise chickens for meat production. It is a large breed and will, on average, take nine to eleven weeks to mature.

Caring for Meat Chickens

Housing

Meat birds tend to be larger than your average layer, so they will require plenty of space. You need to have adequate room both in your coop and in the chicken run for your meat birds. Most meat breeds will grow much faster than the laying birds, so this means space in the coop will free up every so often. However, always ensure that each bird has a minimum of 3 square feet of space in the coop, and if you have a run, the minimum space allowance per bird is at least 8 square feet.

Overcrowded birds tend to spread diseases amongst themselves, fight more, and generally experience more stress. Remember, a healthy bird will give you healthy meat, so even birds that are being raised for meat purposes need to be kept in a healthy and comfortable environment.

A build-up of ammonia in the coop and poor ventilation can also cause problems for your flock. Always ensure that your coop has enough air flowing in and vents to let the air out. Poorly ventilated coops are a breeding ground for disease, and the last thing you want is to get meat from a sick or infected chicken.

Hygiene is key in maintaining your meat birds in a healthy state. Ensure that there is bedding in the coop to help in keeping it clean. Go for bedding such as wood shavings that will absorb and release moisture quickly, leaving the coop dry and odor-free. Clean out the bedding at least monthly to avoid the build-up of manure, which can lead to high ammonia levels in the coop as well as breeding of pests and parasites in the chicken coop.

Meat birds need pretty much the same level of care and maintenance as layers do. Keep them in a clean, comfortable environment, and you will have fewer diseases, bird deaths, and behavioral problems to contend with.

Feeding Meat Birds

Just like any other chickens, your broiler chicks should be started on a starter feed. Starter feed is rich in protein and is specifically formulated to promote growth and proper development in chicks. Starter feed should be given to the baby chicks until they are three weeks old. From this point on, the chicks can be fed grower feed. This feed is meant to promote fast growth and weight gain.

Phase feeding allows your chickens to get all the nutrients they need for the particular age they are at, so it is always important to feed your meet birds age-appropriate food. The key advantage of raising your own meat birds is that you can choose whether to feed them organic or standard feeds. Organic feeds have similar formulations as standard feeds but are typically grown and processed under organic conditions that are certified and approved by relevant regulators. This means that when making organic feeds, companies cannot use anything treated with chemical fertilizers or pesticides, nor any genetically modified compounds. When you choose organic feeds for your meat birds, then you can be sure that the meat you will get once the bird is processed will be free of chemicals or GMO products.

Regardless of whether you choose standard or organic feed for your meat birds, always ensure that the feed you choose meets the basic nutritional requirements. Starter feed should contain at least 22% protein, while grower feeds should contain at least 18% protein. Avoid giving your meat birds layer feed since it contains less protein than broiler feed and may slow down the growth rate of your meat birds.

Once you have the right food, make sure that your meat birds are getting as much food as they need. Meat chickens will, on average, eat more than layers, but since they will mature faster, the average cost will not be that much higher. Have a feeder for every ten birds or so to ensure that each of your chickens has sufficient access to feed. If the feeders are not enough, smaller birds will be bullied and will not get enough food.

Chickens will, on average, drink more water than the feed they consume, so they always need to have access to clean water. You can use waterers in your coop or run to make sure that your meat birds stay hydrated and healthy. Always ensure that the water is clean and free of any dirt or debris.

Meat breeds will generally be ready to be processed at about ten weeks of age, though this can vary from breed to breed. Do not let your meat birds go unprocessed for too long. This is because since they put on a lot of weight very fast if you do not process them at the right time, and they can develop organ failure due to the excess weight they are carrying.

Safety

You are not fattening your meat birds for predators to feed on, so safety should be a top priority when raising meat birds. Your chicken coop should be well-secured and locked up safely at night to keep predators out. Foxes, raccoons, cats, and dogs are all partial to the taste of chicken, so if they get access to your chicken coop, disaster is sure to follow.

Air vents in the chicken coop should be covered with chicken mesh to ensure that only air can get in and out. Check the coop often for rodents, which can hide in the coop bedding and pose a threat to your flock. It is advisable to keep your feed in a different storage area. If you keep chicken feed in the coop, it may attract rodents and other pests to the chicken coop.

Your chicken runs should also be built with the safety of your birds in mind. The fencing should be done using chicken wire or other small mesh fencing material. This will help to keep predators away from your chickens. In some areas, hawks and owls can be a nuisance, so the chicken run may need a cover to stave off flying predators.

Always ensure that you do not let your other pets in the chicken area. Domestic cats and dogs can easily injure chickens, so it is always best to keep them away from your chicken coop and run.

Processing Chickens for Meat

Meat birds will generally be ready to be processed from about ten weeks of age, depending on the particular breed. Before you butcher your bird, make sure you have all the tools that you will need close at hand.

Tools required
- Knives – very sharp, with a 4-inch blade or longer
- Poultry killing cone – available from farm stores
- Bucket

- Clean water – you can run a garden hose to your butchering area
  - Gloves
  - Apron
  - Tarp covered table
  - Scalding water in a huge pot (enough to dunk your bird in)
  - Paper towels
  - Plastic bags or containers for storage

The Butchering Process

I. Once you have captured the bird you want to process, hold it upside down by its feet. In this position, the bird will pass out, making the butchering process easier.

II. Place the chicken in the killing cone

III. Holding the head firmly through the bottom of the killing cone, make a deep firm cut with a sharp knife on the throat.

IV. Once the throat is cut, let the blood drain into the bucket until it is completely drained.

V. Once the blood is drained, remove the killing cone and, still holding the bird upside down by its feet, dunk it into the scalding water.

VI. Make sure the water is hot enough to scald the skin. It should be at least 135F. Swirl the bird around in the scalding water until all the feathers are soaked in the water.

VII. You want the feathers to become loose, but you do not want the chicken skin to tear. Once you pull on a few feathers and they come away easily, remove the bird from the scalding water.

VIII. Hold the bird or suspend it over your bucket and start removing the feathers. You will make much faster progress if you rub your thumb and fingers against the grain of the feathers.

IX. Once you have removed the feathers, rinse the bird with clean water.

X. The next part is preparing the chicken for storage or use.

XI. Hang the chicken up by its feet.

XII. Make a cut from the chicken's groin downwards towards the neck area. As you make the cut, the internal organs will also flow downwards. Cut carefully so that you do not puncture the intestines or any of the other internal organs.

XIII. Once all the organs have fallen (or been pulled) out, rinse out the bird until the water runs clear.

XIV. Finally, you can place the clean bird on your tarp-covered worktable and prep it. You can quarter it by separating it at the joints, or you can store it whole until it is needed.

Chapter 12: Health Care and Maintenance

As far as pets go, chickens are pretty easy going and low maintenance. With a little effort and time on your side, you can have a healthy and happy backyard flock. Most cases of ill health, poor productivity, and death in chickens can be traced back to either a poor diet or unhygienic coop conditions. All this means is that with proper care and maintenance, you should be able to get the best out of your feathery pets.

When it comes to caring for your chickens, the best way to do it is to have scheduled tasks. In this way, nothing gets overlooked, and all your chickens' needs are met on time. So, for most people who raise backyard chickens, having tasks broken down into daily, weekly, and monthly maintenance tasks helps them to keep up with the care of their chickens. This approach will help you in saving time while still giving your pets the best care possible.

Daily Maintenance Tasks

Check the Waterer

Chickens need access to clean water all day every day to stay healthy. You may not need to refill your waterer daily, but you need to ensure that the water is clean and that no dirt or debris has gotten into it. If the water is dirty, replace it with clean water.

Feed the Chickens

Feed your chickens daily. You can choose to use an automatic feeder or gravity feeder that dispenses feed as it is eaten. Always check your feeders daily to ensure that your chickens have enough feed.

Collect Eggs

Collect eggs daily. If you have a large flock of layers, you may need to do this twice or thrice a day to keep the eggs clean and avoid contamination. Leaving eggs in the coop for extended periods may attract predators, and, in some cases, it may cause chickens to start eating their own eggs.

Chicken Watching

Spend a few moments each day observing your flock. This will help you spot any abnormal behavior or signs of illness in your flock. This does not need to take up too much time, and even a few minutes a day can help you stay in touch.

Monthly Maintenance Tasks

Change the Bedding

The bedding in the chicken coop needs to be changed regularly to prevent the accumulation of manure. This is a task that can be done monthly to ensure that your chickens live in a clean and healthy environment. When bedding is not changed often, it can lead to infections and disease.

Clean the Nest Boxes

Just like the rest of the coop, nest boxes need to be kept clean. Remember that this is where your eggs will be laid, and you do not want them contaminated with chicken poop or other kinds of dirt. Change the bedding in the nest box monthly to keep it clean.

Clean your Waterers

At least once a month, make sure that the waterers have been deep-cleaned to remove any contaminants. You can use a mixture of bleach and water to disinfect them completely and then rinse them thoroughly with clean water. Water can easily become a carrier of pathogens and disease-causing germs, so keeping the waterers clean is essential.

Semi-Annual Maintenance Tasks

Deep Cleaning the Coop

Deep cleaning the coop is recommended at least twice a year. This involves washing down all the surfaces in the coop. A mixture of bleach and water can be used to disinfect and sanitize the coop completely. During the deep cleaning process, you can also try sprinkling some diatomaceous earth in the coop. It has been found to help in getting rid of parasites such as lice and mites.

Winter-Proof the Coop

Winter months can be stressful for chickens, and it is important to prepare the coop before winter. In colder months, you may notice that your layers stop laying eggs or lay fewer eggs than they normally would. This is because they do not receive enough daylight to stimulate egg production. Hens typically require a minimum of 16 hours of daylight for egg production. During winter months, egg production will inevitably decrease due to a lack of sufficient daylight. To avoid this situation, put a source of artificial light in the coop during the winter months. This will help to keep your layers productive.

It is also recommended that you add more layers of bedding as winter approaches. Deep litter will help to keep the coop well insulated during the cold season. You may also need heaters for your waterers to keep them from freezing up when temperatures get low.

Keeping Your Flock Healthy

When it comes to health management for your backyard flock, health management will fall into three basic categories:

1. Prevention of disease
2. Early intervention
3. Treatment of disease

Prevention of Disease

The best thing to do for your chickens is not to let them get sick at all. Of course, in some circumstances, this is not always within your control, but in most cases, you can take measures to reduce the risk of diseases. These preventive measures include:

a) Making sure your chicks are vaccinated against common poultry diseases.

b) If your chicks are not vaccinated, using medicated feed may help in boosting the immune system.

c) Maintaining a clean, well-aerated living environment will help to minimize the risks of infections.

d) Providing your chickens with well-balanced, age-appropriate feed to ensure that their basic nutritional needs are being met.

e) Ensuring that your flock has access to clean drinking water at all times.

f) Protecting your flock from extreme conditions such as extreme heat or cold.

g) Keeping your flock safe from predators.

Early Intervention

If you catch signs of poor health early, chances are that the disease will be much easier to treat. This will also prevent the disease from spreading to the entire flock. For this to happen, you need to spend time regularly observing your chickens and taking note of any abnormal behavior.

Here are some warning signs that point to possible underlying conditions that you need to address.

a) Discharge from the nostrils or eyes

b) Droopy wings

c) Lethargy and poor movement/coordination

d) Poor appetite

e) A hen that suddenly stops laying eggs for no apparent reason

f) Weight loss or retarded growth

g) Ruffled feathers

h) Inability to hold their head up

i) Wounds on legs

j) Loss of feathers

When you notice your chicken has any of the above symptoms, it is best to seek help from a vet as soon as possible. You can separate the ill chicken from the rest of the flock to avoid having a disease spread to the rest of the flock.

Treatment of Disease

Getting treatment for any sick chickens is important if you do not want to lose your birds to diseases. Have a vet check on any sick birds and advise on the recommended treatment or next step. Chicken diseases spread pretty fast, and one infected chicken can easily wipe out your entire flock if it is not treated in time.

Common Chicken Diseases

FowlPox

Fowlpox is a common poultry disease. It is spread by mosquitoes, though it also spreads from one chicken to another. Although fowlpox is not necessarily fatal, it can cause death in weak and younger chickens. Fowlpox will usually infect birds for 10-14 days. Some of the symptoms of fowl pox include:

- Comb sores
- White spots on the skin
- Egg production ceases
- Mouth ulcers

Chickens can be vaccinated against fowl pox to minimize the risk of contraction. However, once the birds have contracted the disease, treatment is usually done using supplements of Vitamins A, D, and E. Sick chickens should be fed soft food only until they heal.

Botulism

This disease is caused by food or water contamination. While botulism is not infectious, if your chickens share the same feeder and waterer, they can all get the disease from the contaminated water or feed. Some of the common symptoms of botulism include:

- Feather loss
- Weakness
- Tremors and shakes
- Paralysis which eventually leads to death

If the disease is treated early, the bird can be saved. Some people use a teaspoon of Epsom salts in warm water as a home remedy.

Infectious Bronchitis

This is one of the most common illnesses in backyard flocks. This disease can easily wipe out an entire flock if left untreated. Here are some of the symptoms to be on the lookout for:

- Loss of appetite
- Decreased egg production
- Lethargy
- Eye and nasal discharge
- Misshapen eggs

Ultimately with good care and maintenance, your chickens can live happy and productive lives. Taking care of your pet is not just fulfilling, but it also ensures that you get good quality eggs from your backyard chickens.

Like any other venture, you will learn more and more about the best way to meet your chickens' needs with experience. With time it will be easier for you to identify any problems in the flock and adjust accordingly. Ultimately to raise healthy backyard chickens, you do not need to have a lot of pastureland or spend a lot of money. You can still keep things simple and as natural as possible and raise a productive, happy, and healthy flock.

Conclusion

Nurturing a living thing is probably one of the most rewarding things anyone can ever do. The fulfillment and joy that come from seeing something thrive under your care are invaluable. This is why taking the time to understand how best to take care of your chickens is not just good for your pets but also good for you. Taking the chance to raise your own backyard flock will be so much easier now that you know how to go about it.

Whether you are interested in raising chickens for eggs, for meat, or just the simple pleasure of having an easy-going pet, there are plenty of benefits that come with raising chickens in your backyard. As long as you are willing to dedicate a little time and energy to the care of your chickens, the rewards are bound to outweigh any challenges that you may encounter in the process.

The important thing to remember is that you do not need dozens of chickens to get started. A simple flock of six birds, if well taken care of, can supply you with enough eggs for your family and even surplus that can be sold. Start small and build your flock slowly as you get more knowledgeable about raising chickens, caring for them, and keeping them healthy.

One of the best things about raising chickens is that it is relatively inexpensive. Most of what you need to raise and care for chicken are things that can be easily improvised and made at home. This means that costs should not come between you and the dream of having a flock of healthy backyard chickens to call your own. With a little capital, you will be able to get back most of what you need to get started.

Since you have already taken the first step by equipping yourself with the information and knowledge you need, the next step is simply to start using the knowledge you have acquired and get started setting up for your chickens. The information in this book is timeless and will come in handy whether you choose to start raising chickens today, or sometime in the future.

We hope that you know we have all the tools and information you need to pursue this fulfilling hobby. Finally, if you found the content in this book useful, a review on Amazon is always appreciated.

Here's another book by Dion Rosser that you might like

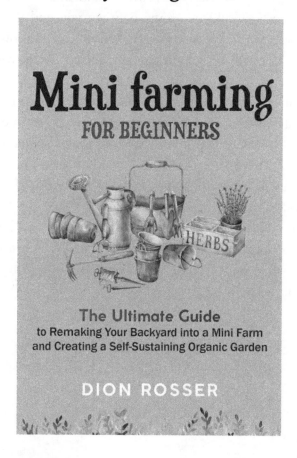

Resources

https://ag.purdue.edu/GMOs/Pages/GMOsandHealth.aspx

https://greatergood.berkeley.edu/article/item/how_modern_life_became_disconnected_from_nature

https://www.fsrmagazine.com/chain-restaurants/whats-americas-most-frequented-restaurant-chain

https://www.downsizinggovernment.org/agriculture/timeline

https://www.ncbi.nlm.nih.gov/books/NBK305168

https://scholarlykitchen.sspnet.org/2020/03/27/a-history-of-panic-buying/

https://www.cnn.com/2003/HEALTH/12/23/madcow.chronology.reut/

https://www.treehugger.com/top-tips-for-the-beginning-homesteader-3016686

http://www.quotehd.com/quotes/words/self%20sufficient

https://www.self-sufficient-farm-living.com/

https://morningchores.com/starting-a-homestead/

https://www.agdaily.com/lifestyle/10-iconic-farming-quotes-history/

https://marketingartfully.com/5-goal-setting-systems/

https://www.workzone.com/blog/project-planning-quotes/

https://www.almanac.com/content/how-build-raised-garden-bed

https://www.thespruce.com/building-a-chicken-coop-3016589

https://www.motherearthnews.com/homesteading-and-livestock/raising-sheep-goats/raising-goats-backyard-farm-ze0z1204zsie

https://completelandscaping.com/much-space-need-fruit-trees/

https://www.blueberrycouncil.org/growing-blueberries/planting-blueberries/

https://www.countryfarm-lifestyles.com/Mini-Farms.html#.XuIkMi2z0Us

https://morningchores.com/assessing-and-planning-homestead/

https://www.primalsurvivor.net/1-acre-tiny-homestead-layouts/

http://www.thebeefsite.com/articles/2415/grazing-small-ruminants-with-cattle/

https://www.diynetwork.com/how-to/outdoors/gardening/manure-compost-

https://en.wikipedia.org/wiki/Cultivar

https://www.thespruce.com/cultivars-vs-varieties-how-do-they-differ-2132281

https://snaped.fns.usda.gov/seasonal-produce-guide

https://gilmour.com/cold-weather-crops

https://www.onegreenplanet.org/lifestyle/perennial-plants/

https://homeguides.sfgate.com/vegetables-grow-yearround-66602.html

https://nellinos.com/the-history-of-the-tomato-in-italy.html

https://www.tropicalpermaculture.com/tropical-vegetables.html

https://www.learningwithexperts.com/gardening/blog/10-flowers-to-grow-with-vegetables

https://www.westernexterminator.com/wasps/what-do-wasps-eat/

https://www.rspb.org.uk/birds-and-wildlife/wildlife-guides/other-garden-wildlife/insects-and-other-invertebrates/flies/hoverfly/

https://www.ufseeds.com/learning/garden-planting-guide/

https://books.google.com/books?id=r5hiDgAAQBAJ&pg=PA11&lpg=PA11&dq=Know+Your+Seed+Varieties+GMO+hybrid+heirloom+cell-fusion&source=bl&ots=nS5wcLmYd0&sig=ACfU3U1TSKgf23MS

NmSLSRQaG9DcK3Q42w&hl=en&sa=X&ved=2ahUKEwjN8da2k
ITqAhXCsZ4KHTa_DPAQ6AEwCXoECAoQAQ#v=onepage&q
=Know%20Your%20Seed%20Varieties%20GMO%20hybrid%20hei
rloom%20cell-fusion&f=false

https://www.offthegridnews.com/how-to-2/best-homesteading-
chickens/

https://www.thehappychickencoop.com/brahma-chicken/

https://morningchores.com/chicken-coop-plans/

https://www.construct101.com/chicken-coop-plans-design-2/

https://www.diyncrafts.com/34313/woodworking/20-free-diy-
chicken-coop-plans-can-build-weekend

https://homesteading.com/how-to-build-a-chicken-coop/

https://104homestead.com/simple-living-kitchen-gadgets/

https://www.homestead.org/food/equip-your-homestead-kitchen-
and-then-make-some-tasty-yogurt/

https://cheesemaking.com/collections/equipment

https://melissaknorris.com/how-to-organize-build-your-homestead-
food-storage-kitchen/

https://apartmentprepper.com/how-to-preserve-meat-without-a-
fridge-2/

https://www.healthline.com/nutrition/fermentation

http://fermentacap.com/how-long-do-fermented-foods-keep/

https://www.culturesforhealth.com/learn/water-kefir/water-kefir-
frequently-asked-questions-faq/

https://www.liveeatlearn.com/homemade-milk-kefir/

https://traditionalcookingschool.com/food-preparation/how-long-
does-kefir-last-aw060/

https://www.jerkyholic.com/how-long-does-beef-jerky-stay-good/

https://www.dummies.com/food-drink/canning/food-preservation-
methods-canning-freezing-and-drying/

http://www.eatingwell.com/article/114109/how-to-pickle-anything-
no-canning-necessary/

https://commonsensehome.com/home-food-
preservation/#4_Freezing

https://commonsensehome.com/root-cellars-101/

https://www.rootwell.com/blogs/root-cellar

https://www.scientificamerican.com/article/experts-organic-milk-lasts-longer/

https://www.mediavillage.com/article/static-branding-vs-organic-branding-uwe-hook-mediabizbloggers/

https://definitions.uslegal.com/f/farmers-market/

http://www.flaginc.org/wp-content/uploads/2013/03/FarmersMarket.pdf

https://www.etsy.com/legal/policy/food-and-edible-items/239327355460

https://www.etsy.com/seller-handbook/article/recipe-for-success-7-tips-for-selling/22506251230

https://www.nolo.com/legal-encyclopedia/starting-home-based-food-business-california.html

https://www.theselc.org/cottage_food_law_summary

https://wwoof.net/

https://wwoofusa.org/how-it-works/be-host

https://4-h.org/

https://www.ffa.org/

https://nifa.usda.gov/cooperative-extension-system

https://www.underatinroof.com/blog/2017/11/15/zmys9ruhc5wis7p5ntzo6idt75zr2i

https://www.grants.gov/

https://www.usda.gov/topics/farming/grants-and-loans

https://www.nal.usda.gov/afsic

https://www.beginningfarmers.org/

https://kidsgardening.org/gardening-basics-garden-maintenance-weeding-mulching-and-fertilizing/

https://homesteading.com/best-homesteading-tools/

https://morningchores.com/low-maintenance-homestead/

https://www.communitychickens.com/hens-stop-laying-zbw2002ztil/

https://www.farmsanctuary.org/wp-content/uploads/2012/06/Animal-Care-Goats.pdf

https://www.farmsanctuary.org/wp-content/uploads/2012/06/Animal-Care-Cattle.pdf

https://www.businessofapps.com/data/youtube-statistics/

https://learn.g2.com/how-much-do-youtubers-make

https://support.patreon.com/hc/en-us/articles/204606315-What-is-Patreon-

https://www.theselfsufficienthomeacre.com/2020/04/how-to-grow-food-in-small-spaces.html

https://www.motherearthnews.com/homesteading-and-livestock/homestead-working-dog-zmaz00aszgoe

https://homesteadsurvivalsite.com/best-dog-breeds-homesteaders/

https://www.thespruce.com/beginners-guide-to-beekeeping-3016857

https://www.motherearthnews.com/organic-gardening/aquaponic-gardening-growing-fish-vegetables-together

https://www.countryliving.com/gardening/garden-ideas/how-to/g1274/how-to-plant-a-vertical-garden/

CPSIA information can be obtained
at www.ICGtesting.com
Printed in the USA
LVHW081451140122
708166LV00012B/92